Everyday Mathematics®

The University of Chicago School Mathematics Project

STUDENT MATH JOURNAL
VOLUME 2

Mc
Graw
Hill
Education

The University of Chicago School Mathematics Project

Max Bell, Director, *Everyday Mathematics* First Edition; James McBride, Director, *Everyday Mathematics* Second Edition; Andy Isaacs, Director, *Everyday Mathematics* Third, CCSS, and Fourth Editions; Amy Dillard, Associate Director, *Everyday Mathematics* Third Edition; Rachel Malpass McCall, Associate Director, *Everyday Mathematics* CCSS and Fourth Editions; Mary Ellen Dairyko, Associate Director, *Everyday Mathematics* Fourth Edition

Authors
Jean Bell, Max Bell, John Bretzlauf, Mary Ellen Dairyko, Amy Dillard, Robert Hartfield, Andy Isaacs, Kathleen Pitvorec, James McBride, Peter Saecker

Fourth Edition Grade 3 Team Leader
Mary Ellen Dairyko

Writers
Lisa J. Bernstein, Camille Bourisaw, Julie Jacobi, Gina Garza-Kling, Cheryl G. Moran, Amanda Louise Ruch, Dolores Strom

Open Response Team
Catherine R. Kelso, Leader; Amanda Louise Ruch, Andy Carter

Differentiation Team
Ava Belisle-Chatterjee, Leader; Martin Gartzman, Barbara Molina, Anne Sommers

Digital Development Team
Carla Agard-Strickland, Leader; John Benson, Gregory Berns-Leone, Juan Camilo Acevedo

Virtual Learning Community
Meg Schleppenbach Bates, Cheryl G. Moran, Margaret Sharkey

Technical Art
Diana Barrie, Senior Artist; Cherry Inthalangsy

UCSMP Editorial
Lila K. S. Goldstein, Senior Editor; Kristen Pasmore, Molly Potnick, Rachel Jacobs

Field Test Coordination
Denise A. Porter

Field Test Teachers
Eric Bachmann, Lisa Bernstein, Rosemary Brockman, Nina Fontana, Erin Gilmore, Monica Geurin, Meaghan Gorzenski, Deena Heller, Lori Howell, Amy Jacobs, Beth Langlois, Sarah Nowak, Lisa Ringgold, Andrea Simari, Renee Simon, Lisa Winters, Kristi Zondervan

Digital Field Test Teachers
Colleen Girard, Michelle Kutanovski, Gina Cipriani, Retonyar Ringold, Catherine Rollings, Julia Schacht, Christine Molina-Rebecca, Monica Diaz de Leon, Tiffany Barnes, Andrea Bonanno-Lersch, Debra Fields, Kellie Johnson, Elyse D'Andrea, Katie Fielden, Jamie Henry, Jill Parisi, Lauren Wolkhamer, Kenecia Moore, Julie Spaite, Sue White, Damaris Miles, Kelly Fitzgerald

Contributors
John Benson, Jeanne Mills DiDomenico, James Flanders, Lila K. S. Goldstein, Funda Gonulates, Allison M. Greer, Catherine R. Kelso, Lorraine Males, Carole Skalinder, John P. Smith III, Stephanie Whitney, Penny Williams, Judith S. Zawojewski

Center for Elementary Mathematics and Science Education Administration
Martin Gartzman, Executive Director; Meri B. Fohran, Jose J. Fragoso, Jr., Regina Littleton, Laurie K. Thrasher

External Reviewers
The *Everyday Mathematics* authors gratefully acknowledge the work of the many scholars and teachers who reviewed plans for this edition. All decisions regarding the content and pedagogy of *Everyday Mathematics* were made by the authors and do not necessarily reflect the views of those listed below.

Elizabeth Babcock, California Academy of Sciences; Arthur J. Baroody, University of Illinois at Urbana-Champaign and University of Denver; Dawn Berk, University of Delaware; Diane J. Briars, Pittsburgh, Pennsylvania; Kathryn B. Chval, University of Missouri–Columbia; Kathleen Cramer, University of Minnesota; Ethan Danahy, Tufts University; Tom de Boor, Grunwald Associates; Louis V. DiBello, University of Illinois at Chicago; Corey Drake, Michigan State University; David Foster, Silicon Valley Mathematics Initiative; Funda Gönülateş, Michigan State University; M. Kathleen Heid, Pennsylvania State University; Natalie Jakucyn, Glenbrook South High School, Glenview, IL; Richard G. Kron, University of Chicago; Richard Lehrer, Vanderbilt University; Susan C. Levine, University of Chicago; Lorraine M. Males, University of Nebraska-Lincoln; Dr. George Mehler, Temple University and Central Bucks School District, Pennsylvania; Kenny Huy Nguyen, North Carolina State University; Mark Oreglia, University of Chicago; Sandra Overcash, Virginia Beach City Public Schools, Virginia; Raedy M. Ping, University of Chicago; Kevin L. Polk, Aveniros LLC; Sarah R. Powell, University of Texas at Austin; Janine T. Remillard, University of Pennsylvania; John P. Smith III, Michigan State University; Mary Kay Stein, University of Pittsburgh; Dale Truding, Arlington Heights District 25, Arlington Heights, Illinois; Judith S. Zawojewski, Illinois Institute of Technology

Note
Many people have contributed to the creation of *Everyday Mathematics*. Visit http://everydaymath.uchicago.edu/authors/ for biographical sketches of *Everyday Mathematics* 4 staff and copyright pages from earlier editions.

www.everydaymath.com

Send all inquiries to:
McGraw-Hill Education
8787 Orion Place
Columbus, OH 43240

ISBN: 978-0-02-143091-8
MHID: 0-02-143091-8

Printed in the United States of America.

II LMN 20

Contents

Unit 6

Unit 7

Multiplication Facts Strategy Logs

Facts Inventory

Activity Sheets

Math Boxes

1 A parallelogram has

_____ pairs of parallel sides.

Draw a parallelogram.

SRB
217

2

4 in.

3 in.

Perimeter = _____ inches

Area = _____ square inches

SRB
174-179

3 Share 3 sandwiches equally among 6 people. How much of a sandwich does each person get?

Draw a picture.

Answer: _____
(unit)

SRB
40, 132

4 Fill in the blanks.

3 ft

9 ft

This is a _____-by-_____ rectangle.

Area = _____
(unit)

Number sentence:

_____ × _____ = _____

SRB
174-179

5 **Writing/Reasoning** Look at Problem 2. Explain how the units used for measuring area are different from the units used for measuring perimeter.

SRB
177

Fractions as Equal Parts

Math Message

Samantha uses a pink fraction circle piece for the whole. She covers it with these 3 pieces:

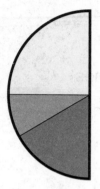

Samantha says each piece shows 1-third of the whole because she covered the pink piece with 3 pieces. Do you agree? Explain.

_____

_____

_____

 Find three equal-size pieces to cover a pink whole.

What color are the equal-size pieces? _____

What fraction of the whole is each piece? _____

Sketch your equal-size pieces on the pink whole:

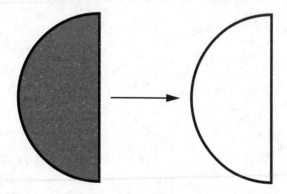

Showing Equal Parts

Exploration A

Use your fraction circles to solve.

1 The pink piece is the whole. Circle the picture that shows 4-fourths of a pink whole.

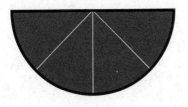   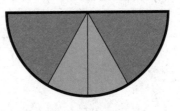

Explain why the picture you circled shows fourths.

_____

2 The yellow piece is the whole. Cover the yellow piece with 1 light green piece and 1 light blue piece. Explain why your pieces do not show 2-halves.

_____

_____

What equal-size pieces can you use to show 2-halves of the yellow whole?

_____

What fraction of the whole is each piece? _____

3 The red piece is the whole. Cover the red piece with 1 yellow piece, 1 orange piece, 1 light blue piece, and 3 light green pieces.

Explain why your pieces do not show 6-sixths. _____

_____

What equal-size pieces can you use to show 6-sixths on the red whole?

_____

What fraction of the whole is each piece? _____

Exploring Wholes

Exploration C

Use your fraction circle pieces to solve.

Example: The orange piece is the whole.

What piece is 1-half of the orange piece?

<u>*A light blue piece*</u>

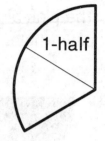

1-half

How do you know? <u>*Two light blue pieces cover an orange piece,*</u>
<u>*so one light blue piece is 1-half of an orange piece.*</u>

1 The red piece is the whole.

What piece is 1-third of the red circle?

_____

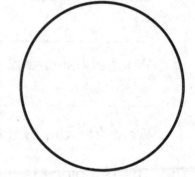

How do you know? _____

_____

Show 1-third on the shape above and label it with a fraction.

Exploring Wholes (continued)

Exploration C

2 The yellow piece is the whole.

What piece is 1-half of a yellow piece?

_____

How do you know? _____

_____

Show 1-half on the shape above and label it.

3 Find the mystery whole. If the yellow piece is 1-half, what piece is the whole?

_____

How do you know?

_____

_____

Draw a picture and label it to show what you did.

Use your fraction circle pieces to figure out each mystery whole.

If a light blue piece is:

4 1-half, what piece is the whole? _____

5 1-third, what piece is the whole? _____

If a dark blue piece is:

6 1-half, what piece is the whole? _____

7 1-fourth, what piece is the whole? _____

8 1-eighth, what piece is the whole? _____

one hundred fifty-three   153

Representing Fractions

Use your fraction circle pieces to help you complete the table. Pay attention to the whole in each problem.

Picture	Words	Number Sentences
The whole is the pink piece.	*2-thirds*	$\frac{1}{3} + \frac{1}{3} = \frac{2}{3}$
① The whole is the orange piece.	3-fourths	
② The whole is the yellow piece.		$\frac{1}{3} + \frac{1}{3} + \frac{1}{3} = \frac{3}{3}$
③ The whole is the pink piece.		
④ The whole is the red circle.		$\frac{1}{8} + \frac{1}{8} + \frac{1}{8} + \frac{1}{8} + \frac{1}{8} = \frac{5}{8}$

Math Boxes

1 Fill in the circles next to the shapes that have at least one right angle.

○ **A.**

○ **B.**

○ **C.**

○ **D.**

SRB
208

2 Circle the shapes that are rhombuses.

SRB
217

3 The following square products have 2 factors that are the same. Fill in the missing factors.

_____ × _____ = 4

_____ × _____ = 25

49 = _____ × _____

64 = _____ × _____

SRB
71

4 Find the area of the shape below. Show your work.

5

2

3

3

Area: _____ square units

SRB
179-180

5 Measure the path to the nearest $\frac{1}{2}$ inch and label each line segment. Then add your measures to find the total distance the caterpillar travels to the leaf.

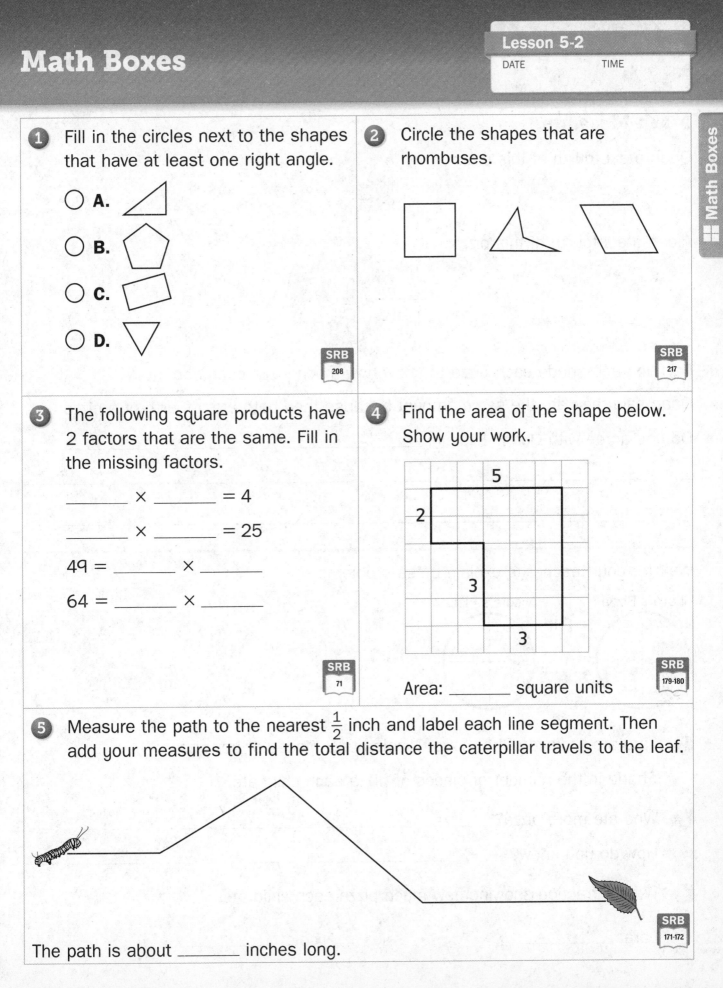

SRB
171-172

The path is about _____ inches long.

Comparing Fractions

Math Message

Quan ate 1-fourth of this pizza:

Aiden ate 1-fourth of this pizza:

Partition and shade each pizza to show how much pizza each boy ate.

Quan said they ate the same amount because they both ate 1-fourth of a pizza.

Do you agree with Quan? Explain. _____

_____

_____

Wait for your teacher to explain these problems.

Lara's Pizza   Nicole's Pizza

1. Lara ate 1 piece of her pizza. Nicole ate 2 pieces of her pizza.

   Shade in the number of pieces of pizza each child ate.

2. Who ate more pizza? _____

   How do you know? _____

3. Write a fraction showing how much pizza each child ate.

   Lara: _____   Nicole: _____

Equivalent Names for Fractions

1 Partition each circle in the name-collection box to show different ways to represent $\frac{1}{2}$. Then add other equivalent fraction names.

1-half

2 **1-fourth**

3 **1-third**

Areas of Rectangles

Fill in the blanks.

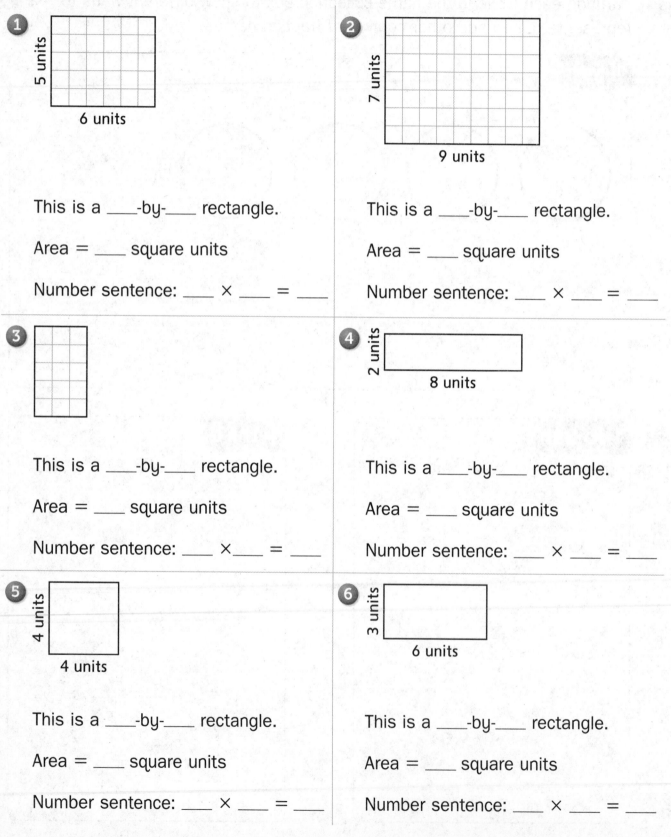

1
5 units
6 units

This is a ____-by-____ rectangle.

Area = ____ square units

Number sentence: ____ × ____ = ____

2
7 units
9 units

This is a ____-by-____ rectangle.

Area = ____ square units

Number sentence: ____ × ____ = ____

3

This is a ____-by-____ rectangle.

Area = ____ square units

Number sentence: ____ ×____ = ____

4
2 units
8 units

This is a ____-by-____ rectangle.

Area = ____ square units

Number sentence: ____ × ____ = ____

5
4 units
4 units

This is a ____-by-____ rectangle.

Area = ____ square units

Number sentence: ____ × ____ = ____

6
3 units
6 units

This is a ____-by-____ rectangle.

Area = ____ square units

Number sentence: ____ × ____ = ____

Math Boxes

1 A rhombus has _____ sides that are all the same length.

Draw a rhombus.

SRB
217

2

4 meters

6 meters

Perimeter = _____ meters

Area = _____ square meters

SRB
174-179

3 Share 3 fruit strips equally among 4 people. You may draw pictures.

Fill in the circles next to ways to show the amount of fruit strip each person could get.

(A) 1-half and 1-fourth

(B) 2-fourths

(C) 1-fourth + 1-fourth + 1-fourth

(D) $\frac{3}{4}$

SRB
40, 132

4 Fill in the blanks.

7 cm

4 cm

This is a _____-by-_____ rectangle.

Area = _____
(unit)

Multiplication number sentence:

_____

SRB
174-179

5 **Writing/Reasoning** Zach wrote this number story to match Problem 4. Zach drew a rectangle and measured its sides. It was 4 cm by 7 cm. What was the area of his rectangle?

Answer: _____
(unit)

Explain how Zach's number story fits Problem 4.

Multiplication Facts Chart

×	0	1	2	3	4	5	6	7	8	9	10
0	0 × 0	1 × 0	2 × 0	3 × 0	4 × 0	5 × 0	6 × 0	7 × 0	8 × 0	9 × 0	10 × 0
1	0 × 1	1 × 1	2 × 1	3 × 1	4 × 1	5 × 1	6 × 1	7 × 1	8 × 1	9 × 1	10 × 1
2	0 × 2	1 × 2	2 × 2	3 × 2	4 × 2	5 × 2	6 × 2	7 × 2	8 × 2	9 × 2	10 × 2
3	0 × 3	1 × 3	2 × 3	3 × 3	4 × 3	5 × 3	6 × 3	7 × 3	8 × 3	9 × 3	10 × 3
4	0 × 4	1 × 4	2 × 4	3 × 4	4 × 4	5 × 4	6 × 4	7 × 4	8 × 4	9 × 4	10 × 4
5	0 × 5	1 × 5	2 × 5	3 × 5	4 × 5	5 × 5	6 × 5	7 × 5	8 × 5	9 × 5	10 × 5
6	0 × 6	1 × 6	2 × 6	3 × 6	4 × 6	5 × 6	6 × 6	7 × 6	8 × 6	9 × 6	10 × 6
7	0 × 7	1 × 7	2 × 7	3 × 7	4 × 7	5 × 7	6 × 7	7 × 7	8 × 7	9 × 7	10 × 7
8	0 × 8	1 × 8	2 × 8	3 × 8	4 × 8	5 × 8	6 × 8	7 × 8	8 × 8	9 × 8	10 × 8
9	0 × 9	1 × 9	2 × 9	3 × 9	4 × 9	5 × 9	6 × 9	7 × 9	8 × 9	9 × 9	10 × 9
10	0 × 10	1 × 10	2 × 10	3 × 10	4 × 10	5 × 10	6 × 10	7 × 10	8 × 10	9 × 10	10 × 10

Applying Adding and Subtracting a Group

For each fact below:

- Think of a helper fact.

- Use your helper fact and either add or subtract a group.

- Write the product for the fact.

Example: $6 \times 8 = ?$

Helper fact: _____ $5 \times 8 = 40$ _____

How I can use it: _ I can add one more group of 8 to 40 _
_ to get $40 + 8 = 48$. _

$6 \times 8 =$ __ 48 __

1 $3 \times 7 = ?$

Helper fact: _____ $2 \times 7 = 14$ _____

How I can use it: _____

_____

$3 \times 7 =$ _____

2 $4 \times 9 = ?$

Helper fact: _____

How I can use it: _____

_____

$4 \times 9 =$ _____

3 $9 \times 8 = ?$

Helper fact: _____

How I can use it: _____

_____

$9 \times 8 =$ _____

Number Story Practice

Solve. Write number models to help keep track of your thinking. Remember to write the units.

1 Savannah earns $5 selling lemonade. Jessica earns double the amount of money that Savannah earns. How much money do they have together?

Number models: _____

_____

Answer: _____
                       (unit)

2 Vijay buys 2 packs of stickers with 8 stickers in each pack. He gives away 3 stickers. How many stickers does he have left?

Number models: _____

_____

Answer: _____
                       (unit)

3 A baker has 10 kilograms of flour. He uses 2 kilograms to bake muffins. He divides the leftover flour into 4 bags. How many kilograms of flour are in each bag?

Number models: _____

_____

Answer: _____
                       (unit)

Try This

4 George is buying wall-to-wall carpeting for two rectangular rooms. One room measures 3 yards by 4 yards. The other room measures 4 yards by 5 yards. How many square yards of carpet does George need to order in all?

Number models: _____

_____

Answer: _____
                       (unit)

Math Boxes

1 Which shapes have at least one pair of parallel sides? Fill in the circles next to the correct answers.

○ **A.**

○ **B.**

○ **C.**

○ **D.**

SRB
209

2 Put an X on the shape that is not a rectangle.

SRB
217

3 Each of the following square products has two factors that are the same. Fill in the missing factors.

9 = _____ × _____

16 = _____ × _____

[   ]   [   ]
× [ ]   × [ ]
_____   _____
3   6   8   1

SRB
71

4 Find the area of the shape below. Show your work.

```
        3
    ┌─────┐
 4  │     │
┌───┘     │
│         │
4│         │
│         │
└─────────┘
     5
```

Area: _____ square units

SRB
179-180

5 Draw a path for the caterpillar to use to get to the leaf. Use a tool to measure the length of your path to the nearest $\frac{1}{2}$ inch.

The path is _____ inches long.

Which tool did you use to measure your path? Why? _____

_____

SRB
171-172

Doubling to Find New Areas

Use the space at the bottom of the page for your sketches.

1. The art table is 2 feet wide and 9 feet long. What is the area of the art table? Draw a sketch of the art table in the empty space below and write a number model to show how you solved the problem.

   Number model: _____ feet × _____ feet = _____ square feet

2. A new art table is 4 feet wide and 9 feet long. Use doubling and your number model from Problem 1 to find the new area. Sketch and write number models to show your work below. (*Hint:* Add to your first drawing.)

   My sketches:

   Number model: _____ feet × _____ feet = _____ square feet

(3) Solve $8 \times 9 = ?$.

Use doubling and your number model from Problem 2 to solve $8 \times 9 = ?$.

You can start by redrawing your sketch of the 4-by-9 table.

Number model: _____ feet × _____ feet = _____ square feet

Library Books Bar Graph

Use the bar graph to solve the problems below.

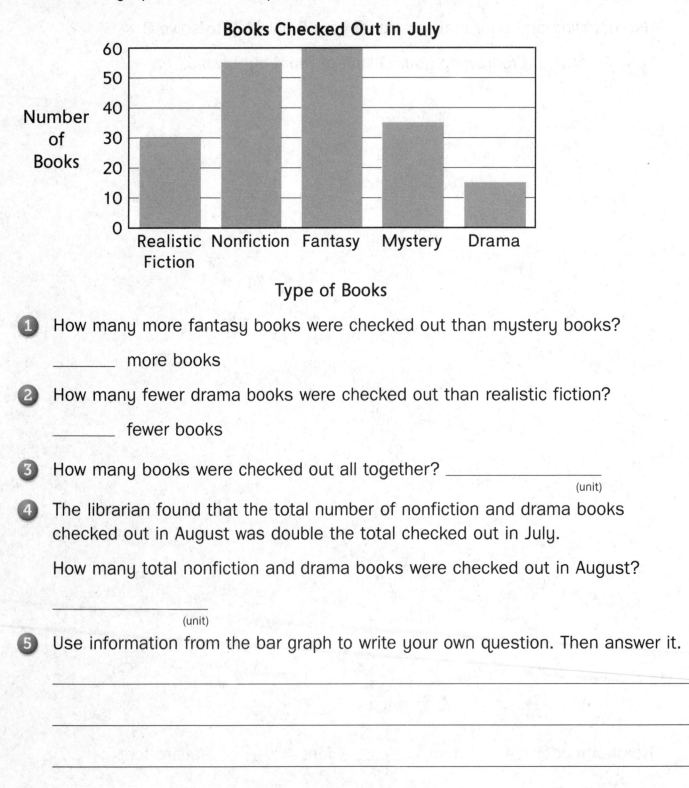

Books Checked Out in July

Number of Books

Type of Books

1 How many more fantasy books were checked out than mystery books?

_____ more books

2 How many fewer drama books were checked out than realistic fiction?

_____ fewer books

3 How many books were checked out all together? _____
                                                                    (unit)

4 The librarian found that the total number of nonfiction and drama books checked out in August was double the total checked out in July.

How many total nonfiction and drama books were checked out in August?

_____
        (unit)

5 Use information from the bar graph to write your own question. Then answer it.

_____

_____

_____

_____

Math Boxes

1 Show how you could use adding or subtracting a group to solve 6 × 8 = ?.

Helper fact:

_____

6 × 8 = _____

SRB 47-48

2 Label each part with a fraction. Shade $\frac{1}{2}$.

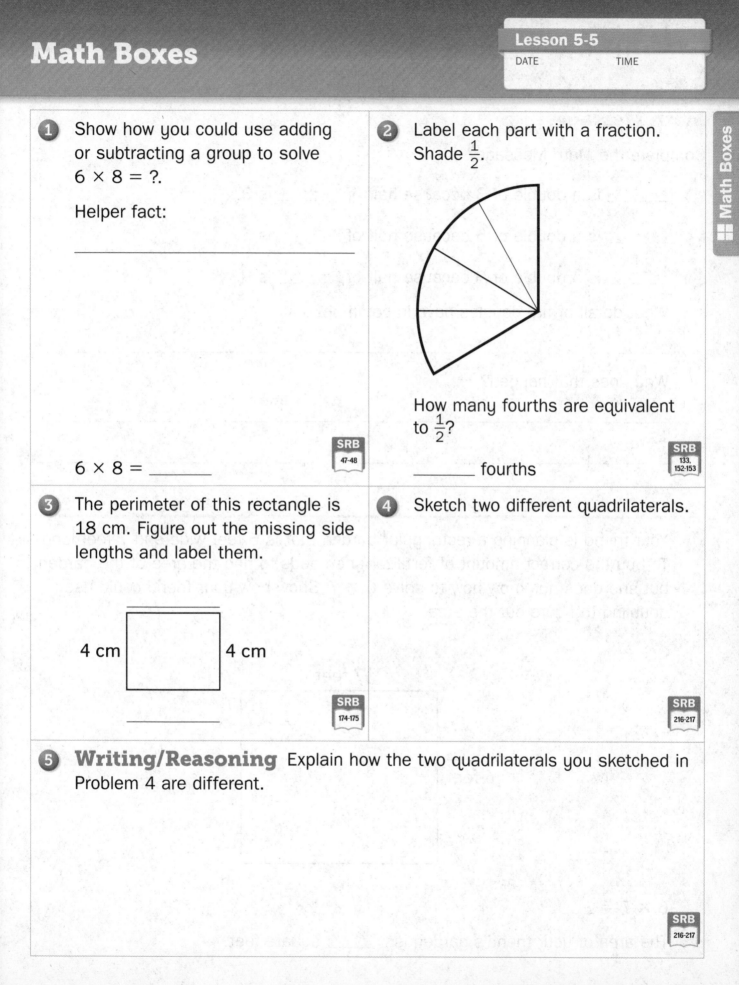

How many fourths are equivalent to $\frac{1}{2}$?

_____ fourths

SRB 133, 152-153

3 The perimeter of this rectangle is 18 cm. Figure out the missing side lengths and label them.

4 cm   4 cm

SRB 174-175

4 Sketch two different quadrilaterals.

SRB 216-217

5 **Writing/Reasoning** Explain how the two quadrilaterals you sketched in Problem 4 are different.

SRB 216-217

Doubling

Math Message

Complete the Math Message.

1 _____ is a double of 3 because half of _____ is 3.

2 _____ is a double of 5 because half of _____ is 5.

3 _____ is a double of 9 because half of _____ is 9.

What do all of the doubles have in common?

_____

Why does that happen?

_____

_____

_____

4 Your friend is planning a rectangular garden that is 6 feet wide and 7 feet long. To buy the correct amount of fertilizer she needs to find the area of the garden, but she does not know how to solve 6 × 7. Show how your friend could use doubling to figure out the area.

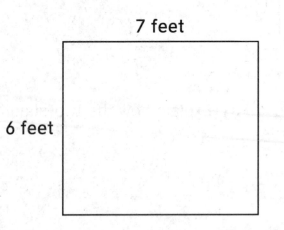

7 feet

6 feet

6 × 7 = _____

The area of your friend's garden is _____ square feet.

Doubling (continued)

5 Your friend changed her mind. Now she wants her garden to be 4 feet wide and 8 feet long. She needs to find the area of this garden, but she does not know how to solve 4 × 8.

a. Use doubling to help her solve 4 × 8. Show your work below. *Hint:* You can start by breaking apart the 4-foot side into 2 feet and 2 feet.

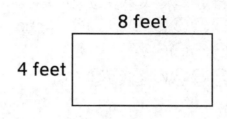

8 feet

4 feet

4 × 8 = _____

The area of your friend's 4-by-8 garden is _____ square feet.

What helper fact did you double to solve 4 × 8?

_____

b. Find a **different** way to use doubling to help your friend find the answer to 4 × 8. Show your work below. *Hint:* You can start by halving the other side of the rectangle.

Break apart the _____-foot side into _____ feet and _____ feet.

8 feet

4 feet

What helper fact did you double to solve 4 × 8?

_____

Creating a Bar Graph from a Picture Graph

Use the information shown in the picture graph to make a bar graph. Remember to add labels and a title to your graph.

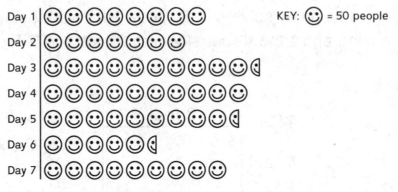

Visitors to Big Bend National Park

Day 1 ☺☺☺☺☺☺☺   KEY: ☺ = 50 people

Day 2 ☺☺☺☺☺☺

Day 3 ☺☺☺☺☺☺☺☺☺(

Day 4 ☺☺☺☺☺☺☺☺☺

Day 5 ☺☺☺☺☺☺☺☺(

Day 6 ☺☺☺☺☺(

Day 7 ☺☺☺☺☺☺☺☺

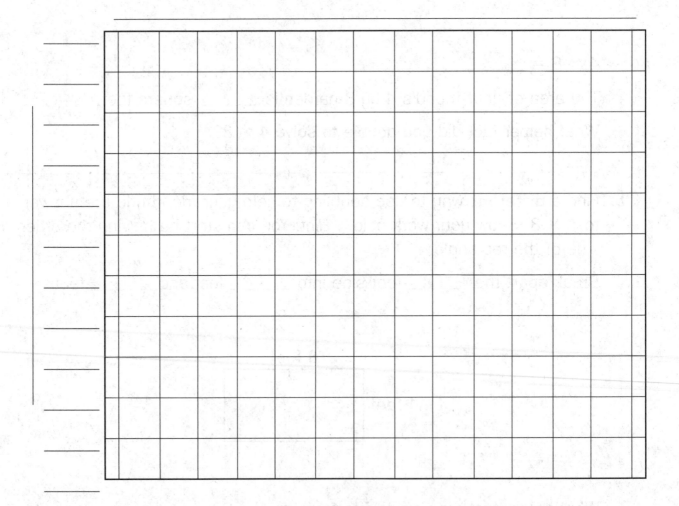

Solving Problems Using Data in Graphs

Big Bend National Park is in southwestern Texas. It contains 801,163 acres of protected wilderness. About 300,000 people visit the park each year.

The data in the graphs on journal page 170 show the number of visitors at Big Bend National Park for one week in August.

Use the information shown in the graphs on journal page 170 to answer the questions below.

1 How many more people visited the park on Day 7 than on Day 6?

_____ people

2 How many fewer people came to Big Bend National Park on Day 1 than on Day 3? _____ fewer people

3 Which 2 days had the most visitors? _____

Which 2 days had the fewest visitors? _____

4 What is the difference between the number of visitors on the 2 days with the most visitors and the 2 days with the fewest visitors? _____ visitors

5 Which graph did you use more often to answer the questions? Why?

_____

_____

Math Boxes

1 Draw a shape with an area of 10 square centimeters.

What is the perimeter of your shape?

_____ centimeters

SRB
176-177

2 Megan visited her friend from 1:15 P.M. to 5:30 P.M. How long was her visit? Fill in the circle next to the correct answer.

○ **A.** 5 hours

○ **B.** 4 hours, 15 minutes

○ **C.** 5 hours, 30 minutes

○ **D.** 4 hours, 45 minutes

SRB
18-19,
187-188

3 One apple has a mass of about 100 grams. What is the mass of 5 apples?

about _____
              (unit)

SRB
114-115

4 Matthew painted $\frac{1}{4}$ of a small picture.

Cherry painted $\frac{1}{4}$ of a large picture. Did they paint the same amount?

_____

SRB
157-158

5 Complete the line plot using the information from the table.

SRB
195

Height of Seedlings

Height of Seedlings	Number of Seedlings
6 cm	2
7 cm	5
8 cm	3
9 cm	7
10 cm	4

centimeters

How many seedlings are 9 centimeters or 10 centimeters tall? _____

How many more seedlings are 9 centimeters than 6 centimeters? _____

Patterns on the Number Grid

-9	-8	-7	-6	-5	-4	-3	-2	-1	0
1	2	3	4	5	6	7	8	9	10
11	12	13	14	15	16	17	18	19	20
21	22	23	24	25	26	27	28	29	30
31	32	33	34	35	36	37	38	39	40
41	42	43	44	45	46	47	48	49	50
51	52	53	54	55	56	57	58	59	60
61	62	63	64	65	66	67	68	69	70
71	72	73	74	75	76	77	78	79	80
81	82	83	84	85	86	87	88	89	90
91	92	93	94	95	96	97	98	99	100
101	102	103	104	105	106	107	108	109	110
111	112	113	114	115	116	117	118	119	120

Describing Patterns in Multiples of 9

Look at the way you circled the multiples of 9 on journal page 173.

① Describe a pattern you see.

_____

_____

_____

_____

② How is this pattern different from the 10s pattern?

_____

_____

_____

_____

③ Explain why the 9s pattern is a diagonal.

_____

_____

_____

_____

Patterns in Even and Odd Products

An **even** number of objects can be divided into two equal parts without any left over.

An **odd** number of objects will have one left over if you try to divide it into two equal groups.

Use your Multiplication/Division Facts Table to help you record examples for each of the columns in the chart below.

2 Even Factors	2 Odd Factors	1 Odd Factor and 1 Even Factor
$2 \times 8 = 16$	$5 \times 9 = 45$	$3 \times 6 = 18$

Use your examples to answer the following questions:

1 If both factors are even numbers, is the product odd or even? _____

2 If both factors are odd numbers, is the product odd or even? _____

3 If one factor is even and the other is odd, is the product odd

   or even? _____

4 Dolores thinks that whenever she multiplies an odd number by an even number the product will be even. She drew a picture of 3×6 to help show this.

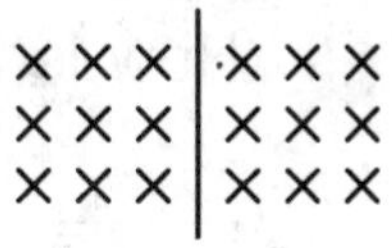

```
× × × | × × ×
× × × | × × ×
× × × | × × ×
```

Explain how Dolores's picture shows why this product is even.

_____

_____

_____

Finding Clock Fractions

Use your fraction circle pieces and toolkit clock to answer the questions.

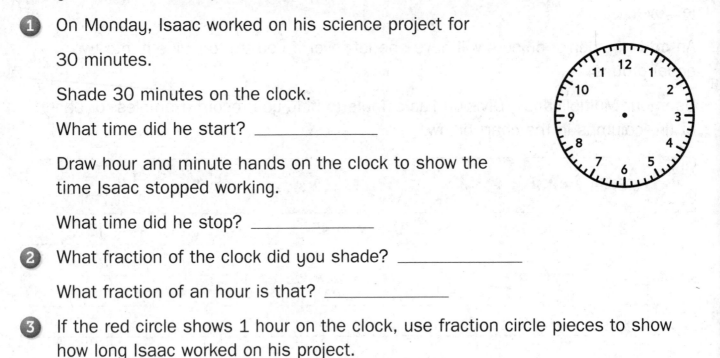

1. On Monday, Isaac worked on his science project for

   30 minutes.

   Shade 30 minutes on the clock.

   What time did he start? _____

   Draw hour and minute hands on the clock to show the
   time Isaac stopped working.

   What time did he stop? _____

2. What fraction of the clock did you shade? _____

   What fraction of an hour is that? _____

3. If the red circle shows 1 hour on the clock, use fraction circle pieces to show
   how long Isaac worked on his project.

   Draw a picture of what you did.

4. How is the clock you shaded like your fraction circle pieces?

   _____

   _____

5. On Tuesday, Isaac worked for 15 minutes. What fraction of an hour did Isaac

   work? You may draw or use your fraction circle pieces. _____

Try This

6. On Wednesday, Isaac worked for 40 minutes. What fraction of an hour did

   Isaac work? _____

Math Boxes

Math Boxes

1 Choose a helper fact for solving
4 × 7 = ?.

Helper fact: _____

Solve by adding or subtracting a
group or by doubling.

4 × 7 = _____

Did you add, subtract, or double?

_____

SRB
44,
47-48

2 Label each part with a fraction.
Shade $\frac{2}{3}$.

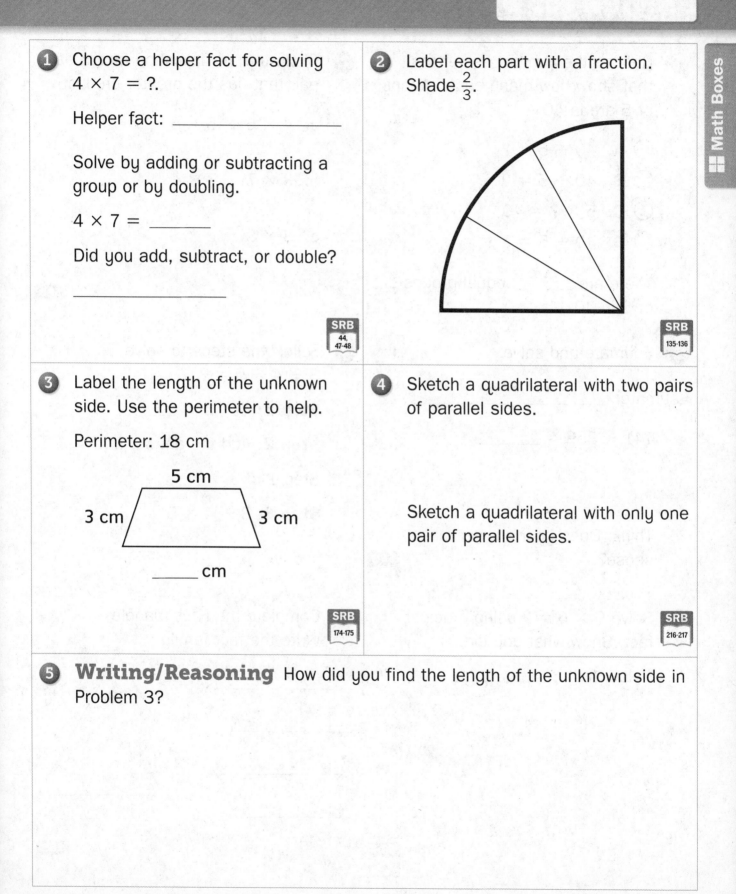

SRB
135-136

3 Label the length of the unknown
side. Use the perimeter to help.

Perimeter: 18 cm

5 cm

3 cm   3 cm

_____ cm

SRB
174-175

4 Sketch a quadrilateral with two pairs
of parallel sides.

Sketch a quadrilateral with only one
pair of parallel sides.

SRB
216-217

5 **Writing/Reasoning** How did you find the length of the unknown side in
Problem 3?

Math Boxes
Preview for Unit 6

1 Choose the number sentences that show how many equal groups of 5 are in 40.

- ○ **A.** $40 + 5 = ?$
- ○ **B.** $40 \div 5 = ?$
- ○ **C.** $5 \times ? = 40$
- ○ **D.** $40 - 5 = ?$

There are _____ equal groups of 5 in 40.

SRB
38-39

2 Circle the multiplication fact in each pair that has the greater product.

a. 4×6 5×6

b. 3×7 7×4

c. 3×8 2×8

SRB
54

3 Estimate and solve.

Unit

Estimate: _____

$741 - 359 =$ _____

Think: Does your answer make sense?

SRB
122-123

4 Follow the steps to solve $3 + 4 \times 5 = ?$.

Step 1: $4 \times 5 =$ _____

Step 2: Add the answer from Step 1 to 3. _____

Step 3: $3 + 4 \times 5 =$ _____

5 Solve $9 \times 6 = ?$ using a helper fact. Show what you did.

Helper fact: _____

SRB
44

6 Complete the Fact Triangle. Write the fact family.

_____

_____

_____

_____

28

×, ÷

7

SRB
53

Math Boxes

1 The top of a table measures 2 feet by 6 feet. Sketch the tabletop. What is its area?

Area: _____ square feet

SRB
179

2 Find the area.

4
4
4
6

Number sentences: _____

_____

_____

Area: _____ square feet

SRB
180

3 Solve. Write another fact next to each using the turn-around rule.

$7 \times 5 =$ ___ _____

$6 \times 10 =$ ___ _____

$9 \times 2 =$ ___ _____

$8 \times 3 =$ ___ _____

SRB
45

4 Draw two different quadrilaterals that are **not rhombuses.**

SRB
216-217

5 **Writing/Reasoning** Look at how you solved Problem 1. How can you use 2×6 to help you solve 4×6?

SRB
49

Multiplication Facts Chart

×	0	1	2	3	4	5	6	7	8	9	10
0	0 × 0	1 × 0	2 × 0	3 × 0	4 × 0	5 × 0	6 × 0	7 × 0	8 × 0	9 × 0	10 × 0
1	0 × 1	1 × 1	2 × 1	3 × 1	4 × 1	5 × 1	6 × 1	7 × 1	8 × 1	9 × 1	10 × 1
2	0 × 2	1 × 2	2 × 2	3 × 2	4 × 2	5 × 2	6 × 2	7 × 2	8 × 2	9 × 2	10 × 2
3	0 × 3	1 × 3	2 × 3	3 × 3	4 × 3	5 × 3	6 × 3	7 × 3	8 × 3	9 × 3	10 × 3
4	0 × 4	1 × 4	2 × 4	3 × 4	4 × 4	5 × 4	6 × 4	7 × 4	8 × 4	9 × 4	10 × 4
5	0 × 5	1 × 5	2 × 5	3 × 5	4 × 5	5 × 5	6 × 5	7 × 5	8 × 5	9 × 5	10 × 5
6	0 × 6	1 × 6	2 × 6	3 × 6	4 × 6	5 × 6	6 × 6	7 × 6	8 × 6	9 × 6	10 × 6
7	0 × 7	1 × 7	2 × 7	3 × 7	4 × 7	5 × 7	6 × 7	7 × 7	8 × 7	9 × 7	10 × 7
8	0 × 8	1 × 8	2 × 8	3 × 8	4 × 8	5 × 8	6 × 8	7 × 8	8 × 8	9 × 8	10 × 8
9	0 × 9	1 × 9	2 × 9	3 × 9	4 × 9	5 × 9	6 × 9	7 × 9	8 × 9	9 × 9	10 × 9
10	0 × 10	1 × 10	2 × 10	3 × 10	4 × 10	5 × 10	6 × 10	7 × 10	8 × 10	9 × 10	10 × 10

Exploring Near Squares

Add or subtract a group from a multiplication square helper fact to solve each near-squares fact.

Use drawings, words, or numbers to explain your thinking.

 1 Near-squares fact: 3 × 4 = ?

Multiplication square helper fact: _____

How I solved it:

3 × 4 = _____

2 Near-squares fact: 7 × 6 = ?

Multiplication square helper fact: _____

How I solved it:

7 × 6 = _____

3 Near-squares fact: 9 × 8 = ?

Multiplication square helper fact: _____

How I solved it:

9 × 8 = _____

4 Do your own.

Near-squares fact: _____ × _____ = _____

Multiplication square helper fact: _____ × _____ = _____

How I solved it:

Subtraction Practice

Use the expand-and-trade method to solve.

Remember to compare your answers to your estimates to check whether they make sense.

Unit

books

SRB
119

---

1 **Example:**

Estimate:

$$300 - 100 = 200$$

$$
\begin{array}{r}
307 \rightarrow 300 + 0 + 7 \\
- 129 \rightarrow 100 + 20 + 9 \\
\hline
100 + 70 + 8 = 178
\end{array}
$$

(traded: 200 100 17, with 90 above)

$307 - 129 = \underline{178}$

---

2 Estimate:

_____

$580 - 446 = $ _____

---

3 Estimate:

_____

$704 - 285 = $ _____

---

4 Estimate:

_____

$510 - 355 = $ _____

Making Sense of a Problem

There are 10 children on Maurice's baseball team. The coach gives each child 2 granola bars from a package of 24 bars. The coach gets the leftover granola bars. How many granola bars does Maurice's coach get?

What do you need to find out?

Use words or pictures to show what you know about the problem and how to solve it.

Math Boxes

1 Draw a shape with an area of 20 square centimeters.

SRB 176-177

What is the perimeter of your shape? _____ centimeters

2 Sanjay began working on his school project at 6:10 P.M. It took him 2 hours and 30 minutes to finish the project. What time did he finish?

○ **A.** 7:40 P.M.

○ **B.** 8:10 P.M.

○ **C.** 8:40 P.M.

○ **D.** 3:40 P.M.

SRB 18-19, 187-188

3 A child's filled backpack has a mass of about 7 kilograms. What is the mass of 3 such backpacks?

about _____
                          (unit)

Write a number model that fits the story.

_____

SRB 30

4 Sam ate $\frac{2}{3}$ of a small muffin. Maria ate $\frac{2}{3}$ of a large muffin. Did they eat the same amount? _____

Draw a picture to explain.

SRB 157-158

5 Complete the line plot using the information from the table.

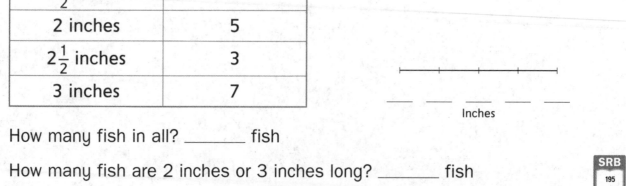

Length of Fish	Number of Fish
1 inch	4
$1\frac{1}{2}$ inches	1
2 inches	5
$2\frac{1}{2}$ inches	3
3 inches	7

Length of Fish

Inches

How many fish in all? _____ fish

How many fish are 2 inches or 3 inches long? _____ fish

SRB 195

Breaking Apart in Different Ways

Math Message

The marching band is planning their next show. They begin by forming an array with 7 rows and 6 marchers in each row. Then they separate into two smaller arrays that still have 6 marchers in each row. Use centimeter cubes to show one way they could do this. Record your thinking by shading in squares on one of the the grids below. Then repeat, showing a *different* way the band could do this.

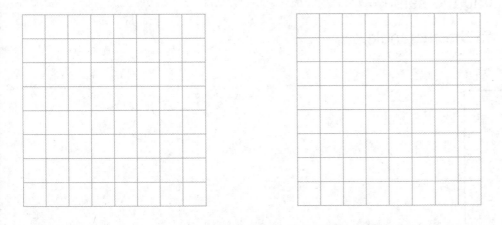

1. You have a rectangular garden that is 7 feet wide and 8 feet long. You decide to plant flowers in one section and vegetables in another. Sketch at least two different ways you can partition, or divide, your garden into two rectangular sections. Label the side lengths of each of your new rectangles.

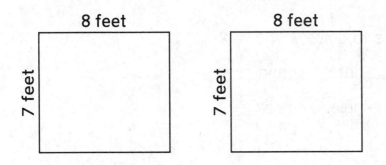

Write a number model using easier helper facts for one of your ways.

$7 \times 8 = \underline{\quad} \times \underline{\quad} + \underline{\quad} \times \underline{\quad}$

$7 \times 8 = \underline{\quad}$

2) Your friend wants to solve 8 × 9. You suggest that she imagine a garden that is 8 feet wide and 9 feet long. Help her break 8 × 9 into two smaller helper facts using the rectangular garden.

Show one way to break apart the 8-by-9 foot garden.

I will break apart the factor ____ into ____ and ____.

The factor ____ will stay the same.

Show what you did on the rectangle.

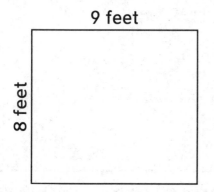

9 feet

8 feet

Helper facts that match the areas of the smaller rectangles:

_____ × _____ = _____   and   _____ × _____ = _____

8 × 9 = _____

3) Break apart one of the factors to solve 7 × 6 = ?.

I will break apart the factor ____ into ____ and ____.

Show what you did on the rectangle.

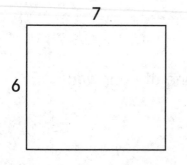

7

6

7 × 6 = _____

Math Boxes

1 A rug is 3 feet by 4 feet. What is its area?

4 feet

3 feet

Area of 3-by-4 rug: _____ sq ft

Now find the area of a rug that is 3 feet by 8 feet. Add to the rug above to show doubling.

Area of 3-by-8 rug: _____ sq ft

SRB
49, 179

2 Find the area. Write number sentences to show your strategy.

8

6   6

2   2

Number sentences:

Area: _____ square units

SRB
180

3 Solve. Write a fact next to each problem using the turn-around rule.

___ = 9 × 3 _____

___ = 2 × 8 _____

___ = 9 × 5 _____

SRB
45

___ = 7 × 10 _____

4 Draw a quadrilateral that is not a rhombus and not a rectangle.

SRB
216-217

5 **Writing/Reasoning** Explain how you found the area of the shape in Problem 2.

Math Boxes
Preview for Unit 6

(1) There are 6 bagels in each bag. Shana has 36 bagels in all. How many bags of bagels does Shana have? Fill in the circle next to the correct answer.

- ○ **A.** 4 bags
- ○ **B.** 5 bags
- ○ **C.** 6 bags
- ○ **D.** 10 bags

SRB
38-39

(2) Circle the multiplication fact in each pair that has the greater product.

a. 6×6 7×7

b. 9×8 9×4

c. 1×8 1×7

SRB
54

(3) Estimate and then solve.

Unit

Estimate: _____

$$\begin{array}{r} 8\ 3\ 2 \\ -\ 2\ 6\ 8 \\ \hline \end{array}$$

Think: Does your answer make sense?

SRB
122-123

(4) Follow the steps to solve $40 - 3 \times 5 = ?$.

Step 1: $3 \times 5 =$ _____

Step 2: Subtract the answer for Step 1 from 40. _____

Step 3: $40 - 3 \times 5 =$ _____

(5) Solve $4 \times 8 = ?$ using a helper fact. Show what you did.

Helper fact: _____

SRB
44

(6) Complete the Fact Triangle. Write the fact family.

_____

_____

_____

_____

42

$\times, \div$

6

_____

SRB
53

Subtraction Strategies

Fill in the unit box. Solve Problem 1 using expand-and-trade subtraction.

1 Estimate: _____

```
  2 3 1
- 1 7 4
```

Unit

Is your answer reasonable? How do you know? _____

_____

2 Solve Problem 2 using trade-first subtraction.

Estimate: _____

```
  4 0 6
- 3 8 9
```

Does your answer make sense? How do you know?

_____

Trade-First Subtraction

Fill in the unit box. Then use trade-first subtraction to solve.
Check your answers.

Unit

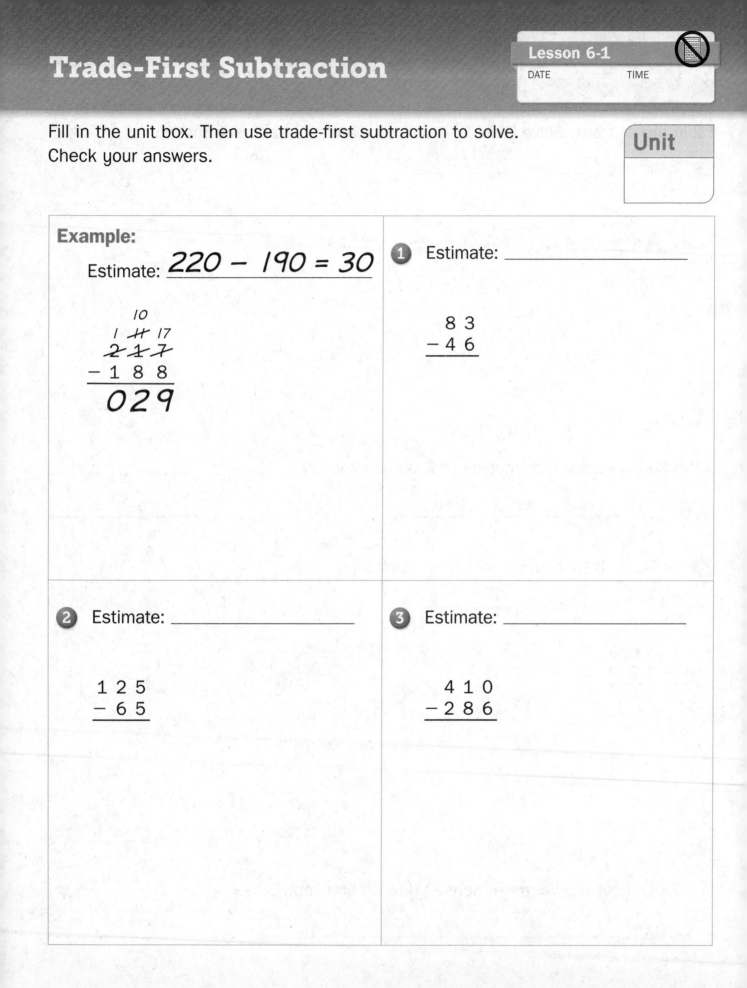

Example:

Estimate: $220 - 190 = 30$

$$\begin{array}{r} {\scriptstyle 10} \\ {\scriptstyle 1\ \not{11}\ 17} \\ \not{2}\ \not{1}\ \not{7} \\ -\ 1\ 8\ 8 \\ \hline 029 \end{array}$$

① Estimate: _____

$$\begin{array}{r} 8\ 3 \\ -\ 4\ 6 \\ \hline \end{array}$$

② Estimate: _____

$$\begin{array}{r} 1\ 2\ 5 \\ -\ 6\ 5 \\ \hline \end{array}$$

③ Estimate: _____

$$\begin{array}{r} 4\ 1\ 0 \\ -\ 2\ 8\ 6 \\ \hline \end{array}$$

Math Boxes

1 The perimeter of this square is

_____ centimeters.

4 cm

SRB
174-175

2 It is 7:40 A.M. Delilah's bus arrives at 7:53 A.M. It takes her 10 minutes to get to the bus stop. Will she catch her bus?

_____

What time will she get to the bus stop?

_____

SRB
18-19

3 Measure the line segment to the nearest $\frac{1}{2}$ inch.

_____

about _____ inch(es)

Draw a line segment that is $1\frac{1}{2}$ inches long.

SRB
171-172

4 Use the rule to fill in the missing numbers. Write *in* and *out* numbers in the last row.

in
↓

Rule

× 3

↓
out

in	out
2	
4	
6	

SRB
73-74

5 **Writing/Reasoning** Find the area of the square in Problem 1. Explain how you found the area and the perimeter.

Area = _____ square centimeters

SRB
176-179

Multiplication Riddle

Math Message

Follow the directions to solve the riddle.

Riddle: What animal is good at hitting a baseball?

Solve each fact below. Match each product to a letter in the key. Then write the letter on the line above the fact to find the answer to the riddle.

Key							
Letter:	A	B	E	L	M	S	T
Product:	0–15	16–30	31–45	46–60	61–75	76–90	91–100

$$\overline{}\ \overline{}\ \overline{}\ \overline{}!$$
3 × 3   4 × 4   2 × 2   10 × 10

Explain the answer to the riddle to a partner.

Math Boxes

1 Fill in the circles next to ways to break 8 × 6 into two smaller multiplication facts.

Ⓐ 4 × 6, 4 × 6

Ⓑ 8 × 3, 8 × 3

Ⓒ 5 × 6, 6 × 6

Ⓓ 2 × 6, 6 × 6

Solve.

8 × 6 = _____

SRB
49, 51

2 Find the area. You may use the doubling or break-apart strategy.

7 ft

4 ft

Area = _____
                        (unit)

SRB
49, 51,
179

3 Fill in the missing numbers.

in

Rule

× 3

out

in	out
2	
5	
	9
	30

SRB
73-74

4 Record a fraction equivalent to $\frac{1}{2}$ that represents the shaded parts of the circle on the right.

$\frac{1}{2} = \dfrac{\boxed{}}{\boxed{}}$

SRB
150, 153

5 Complete.

× 4

÷ 2

1   4

SRB
72-73

Identifying Facts That Fit Strategies

 1 Look at the class posters. In the boxes below, record facts that you could solve using each strategy.

Adding a group	Doubling
Subtracting a group	**Near squares**

2 How are the facts in the doubling box alike?

_____

_____

3 What do you notice about the facts in the near-squares box?

_____

_____

4 What do you notice about the facts in the adding-a-group box and the subtracting-a-group box?

_____

_____

Math Boxes

1 Find the perimeter and the area of the rectangle.

7 meters

5 meters

Perimeter: _____
                                    (unit)

Area: _____
                        (unit)

SRB
174-175,
179

2 It is 2:57 P.M.

It takes Patrick's dad 40 minutes to drive to school. School ends at 3:30 P.M. Will Patrick's dad be late?

_____

What time will he arrive at school?

_____

SRB
18-19

3 Measure the line segment to the nearest $\frac{1}{2}$ inch.

_____

about _____ inches

Draw a line segment that is $2\frac{1}{2}$ inches long.

SRB
171-172

4 Use the rule to fill in the missing numbers. Write *in* and *out* numbers in the last row.

in

Rule

× 1

out

in	out
1	
7	
4	

SRB
73-74

5 **Writing/Reasoning** What patterns do you notice in the *in* and *out* numbers in Problem 4? Explain.

SRB
46

Math Boxes

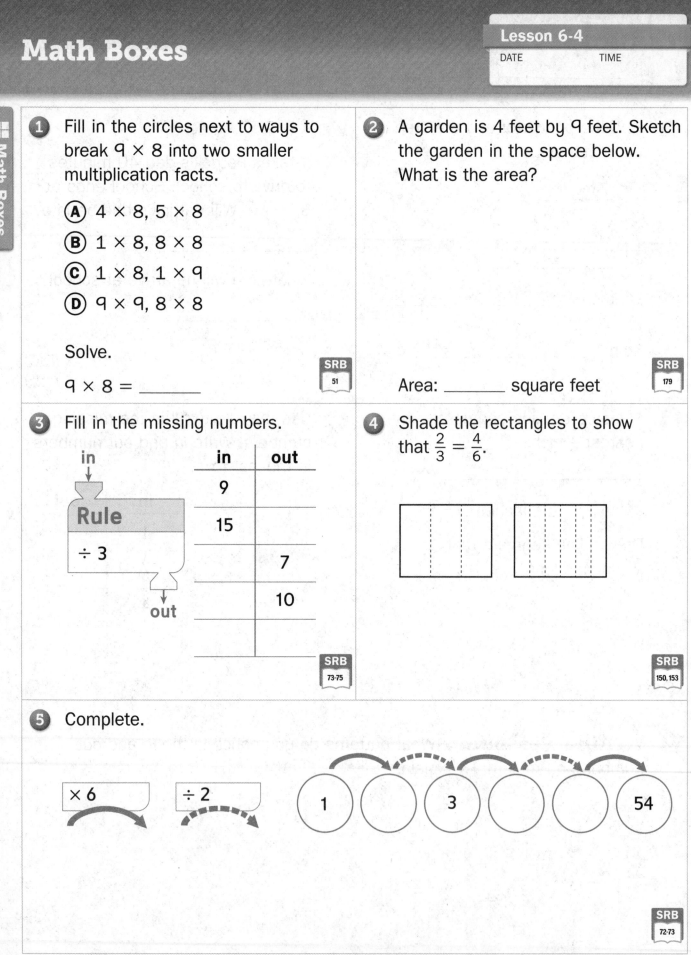

Math Boxes

① Fill in the circles next to ways to
break 9 × 8 into two smaller
multiplication facts.

Ⓐ 4 × 8, 5 × 8

Ⓑ 1 × 8, 8 × 8

Ⓒ 1 × 8, 1 × 9

Ⓓ 9 × 9, 8 × 8

Solve.

9 × 8 = _____

SRB
51

② A garden is 4 feet by 9 feet. Sketch
the garden in the space below.
What is the area?

Area: _____ square feet

SRB
179

③ Fill in the missing numbers.

in
↓

Rule

÷ 3

↓
out

in	out
9	
15	
	7
	10

SRB
73-75

④ Shade the rectangles to show
that $\frac{2}{3} = \frac{4}{6}$.

SRB
150, 153

⑤ Complete.

× 6   ÷ 2   1   3   54

SRB
72-73

For each problem, use straws and twist ties to make the shape. Then draw a picture of your shape.

 1 Make a rhombus that is not a square.

Drawing:

2 Make a quadrilateral that is both a parallelogram and a rhombus.

Drawing:

3 Make a different quadrilateral.

Drawing:

 4 Write at least three attributes of the shape you drew in Problem 3.

_____

_____

Exploration C: Comparing Polygon Measurements

Follow the directions on Activity Card 75.

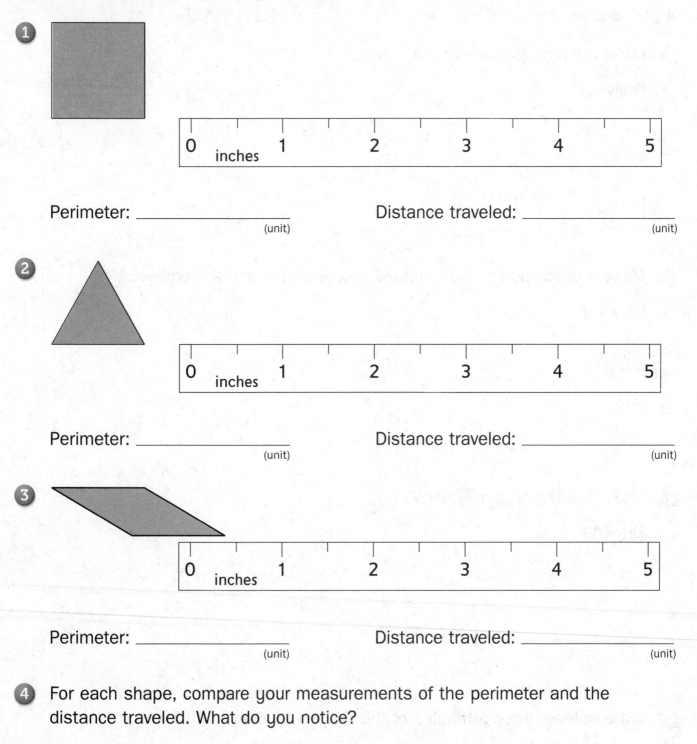

1

0 inches 1 2 3 4 5

Perimeter: _____   Distance traveled: _____
                    (unit)   (unit)

2

0 inches 1 2 3 4 5

Perimeter: _____   Distance traveled: _____
                    (unit)   (unit)

3

0 inches 1 2 3 4 5

Perimeter: _____   Distance traveled: _____
                    (unit)   (unit)

4 For each shape, compare your measurements of the perimeter and the distance traveled. What do you notice?

_____

_____

Math Boxes

■ Math Boxes

1 Solve.

Estimate: _____

Unit

$$\begin{array}{r} 1\ 9\ 3 \\ -\ \ \ 4\ 9 \\ \hline \end{array}$$

Think: Does my answer make sense?

SRB
119,
122-123

2 Draw a quadrilateral that is a parallelogram.

SRB
216-217

3 The school has 385 water bottles for field day. The third grade brings 5 packs of 8 water bottles. How many water bottles are there now?

Number models:_____

_____

Answer: _____ water bottles

SRB
30-31

4 The circle is the whole. Label each part with a fraction.

What fraction of the circle is

shaded? _____

SRB
132-134

5 **Writing/Reasoning** How can you check your answer for Problem 1?

SRB
103

More Number Stories

For each number story:

- Complete the diagram. Use a letter to represent the unknown amount.

- Write a number model.

- Solve the number story. You may draw a picture to help.

- Write your answer with the unit. Think: Does my answer make sense?

1 Adele has 50 stickers to put into 5 gift bags. She wants the same number of stickers in each bag. How many stickers should she put in each bag?

Letter and what it represents: _____

gift bags	stickers per gift bag	stickers in all

_____
(number model with letter)

Answer: _____
                                         (unit)

2 There are 48 third graders. The gym teacher groups them into teams of 6. How many teams are there?

Letter and what it represents: _____

teams	third graders per team	third graders in all

_____
(number model with letter)

Answer: _____
                                         (unit)

For Problems 3 and 4, fill in the diagrams to help organize the information from the stories.

3 There are 7 boxes of golf balls. Each box has the same number of balls. There are 63 total golf balls. How many golf balls are in each box?

Letter and what it represents: _____

_____	_____ **per** _____	_____ **in all**

_____
(number model with letter)

Answer: _____
(unit)

Try This

4 A package of notebook paper contains 80 sheets of paper. Your teacher has 4 packages of paper. How many sheets of paper does your teacher have?

Letter and what it represents: _____

_____	_____ **per** _____	_____ **in all**

_____
(number model with letter)

Answer: _____
(unit)

Math Boxes

① I have four corners. I have at least two parallel sides. I have no right angles. What could I be?

Fill in the circles next to the possible answers.

Ⓐ trapezoid

Ⓑ square

Ⓒ hexagon

Ⓓ parallelogram

SRB
216-217

② Divide this shape into two rectangles to find its area.

SRB
180

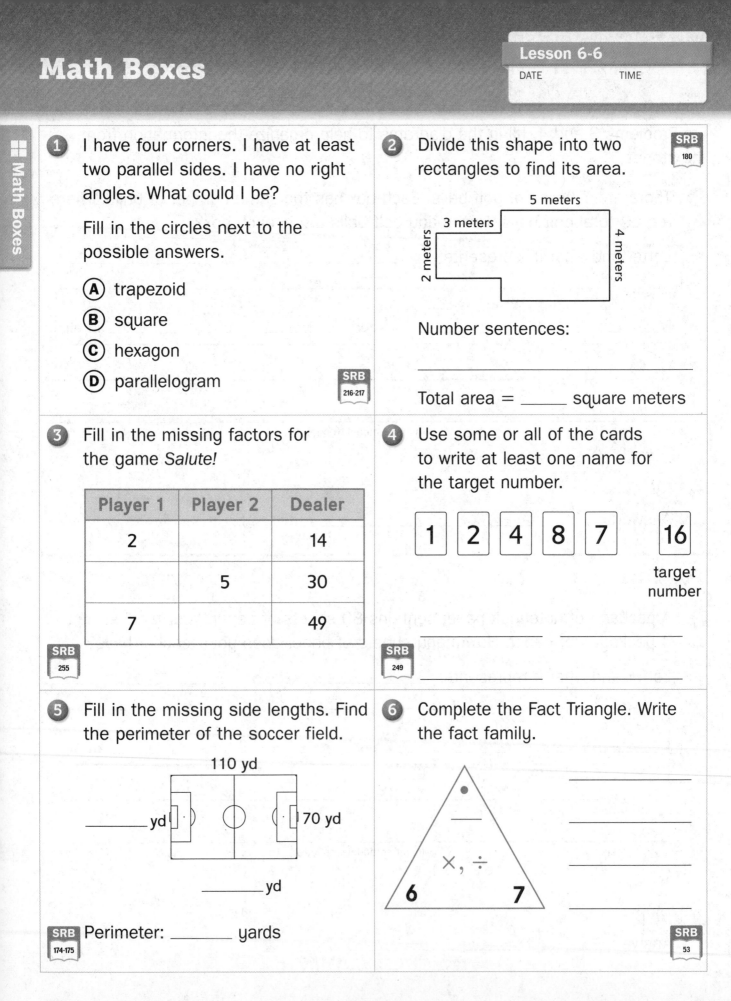

5 meters

3 meters

2 meters

4 meters

Number sentences:

_____

Total area = _____ square meters

③ Fill in the missing factors for the game *Salute!*

Player 1	Player 2	Dealer
2		14
	5	30
7		49

SRB
255

④ Use some or all of the cards to write at least one name for the target number.

| 1 | 2 | 4 | 8 | 7 | 16 |

target number

_____   _____

_____   _____

SRB
249

⑤ Fill in the missing side lengths. Find the perimeter of the soccer field.

110 yd

_____ yd

70 yd

_____ yd

SRB
174-175

Perimeter: _____ yards

⑥ Complete the Fact Triangle. Write the fact family.

×, ÷

6   7

_____

_____

_____

_____

SRB
53

Multiplying with Larger Factors

Sadie uses a multiplication strategy to solve 5 × 12.

Sadie's strategy:

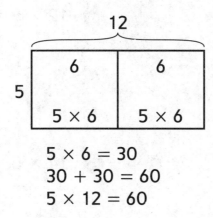

$5 \times 6 = 30$

$30 + 30 = 60$

$5 \times 12 = 60$

 Explain Sadie's strategy.

_____

_____

_____

_____

② Use a different strategy to solve 5 × 12. Use drawings, words, or numbers to show your thinking.

More Multiplying with Larger Factors

Solve. Use pictures, words, or numbers to show your strategies.
Look at the Fact Strategy Wall for help.

1 6 × 15 = _____

2 12 × 3 = _____

3 _____ = 11 × 4

4 14 × 5 = _____

5 _____ = 16 × 7

6 _____ = 17 × 4

Try This

7 Keisha is buying fabric to make costumes for the school play. She needs at least 50 square yards. She has a 4-yard by 12-yard piece of fabric. Does she have enough fabric?

_____

Show your thinking.

Estimating and Measuring Mass

① Record at least one benchmark object for each mass.

1 gram	500 grams	1 kilogram

② Estimate the masses of objects in your classroom. Use the benchmark items to help.

- Write the names of the objects and your estimates in the table below.

- Use a pan balance and standard masses to find the actual masses of your objects. Record your work in the table. Write the units.

Name of Object	Estimated Mass	Actual Mass

③ How do benchmark objects help you estimate the masses of other objects?

_____

_____

Math Boxes

Math Boxes

① Solve.

Estimate: _____

	3	7	1
−		5	7

Unit

Think: Does my answer make sense?

SRB
103, 119, 122-123

② Draw a quadrilateral that is not a parallelogram.

SRB
216-217

③ The basketball team has $125. Team T-shirts cost $10 each. How much money will the team have left over after buying 9 T-shirts?

Number models: _____

_____

Answer: $_____

SRB
30-31

④ The circle is the whole. Label each part with a fraction.

What fraction of the whole circle is *not* shaded? _____

SRB
132-134

⑤ **Writing/Reasoning** Zeke thinks that the unshaded part in Problem 4 is $\frac{3}{5}$. Do you agree? Explain.

Number Sentences with Parentheses

Complete these number sentences.

1 _____ = 18 − (9 + 5)

2 (75 − 29) + 5 = _____

3 _____ = 8 + (9 × 3)

4 36 + (15 + 3) = _____

5 50 − (10 ÷ 2) = _____

6 (15 + 5) ÷ 5 = _____

Draw parentheses to make each number sentence true.

7 20 − 10 + 4 = 6

8 20 − 10 + 4 = 14

9 100 − 21 + 10 = 69

10 100 − 21 + 10 = 89

11 Amanda and Jerry solved this number sentence: 32 − (5 + 7) = ?
Amanda says the answer is 34, and Jerry says the answer is 20.

Who remembered what parentheses mean? Explain.

_____

_____

_____

Math Boxes
Preview for Unit 7

Math Boxes

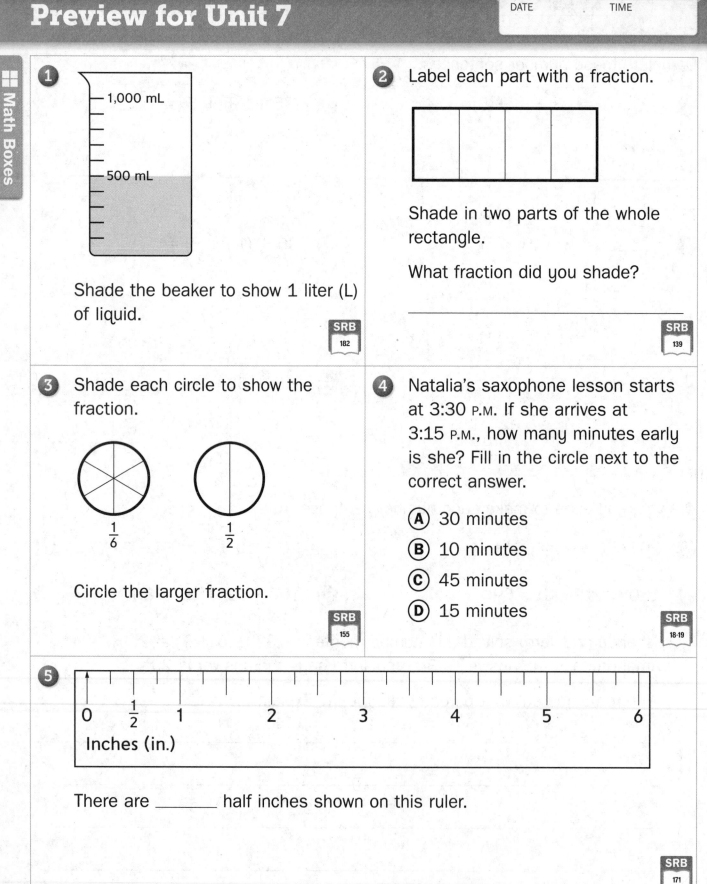

1

1,000 mL

500 mL

Shade the beaker to show 1 liter (L) of liquid.

SRB 182

2 Label each part with a fraction.

Shade in two parts of the whole rectangle.

What fraction did you shade?

_____

SRB 139

3 Shade each circle to show the fraction.

$\frac{1}{6}$ $\frac{1}{2}$

Circle the larger fraction.

SRB 155

4 Natalia's saxophone lesson starts at 3:30 P.M. If she arrives at 3:15 P.M., how many minutes early is she? Fill in the circle next to the correct answer.

(A) 30 minutes

(B) 10 minutes

(C) 45 minutes

(D) 15 minutes

SRB 18-19

5

0    $\frac{1}{2}$ 1   2   3   4   5   6

Inches (in.)

There are _____ half inches shown on this ruler.

SRB 171

Connecting a Number Story and a Number Model

Quincy played 3 soccer games. In each game, he scored 2 goals. How many more goals does Quincy need to score a total of 10 goals?

Use this number model to solve the problem. G represents the number of goals Quincy needs to score.

$10 - (3 \times 2) = G$

$10 - (3 \times 2) =$ _____ goals

Explain to a partner how the number model fits the number story.

Math Boxes

Math Boxes

---

1 36 markers are shared among 6 children. How many markers does each child get?

Number of children	Number of markers per child	Number of markers in all

_____
(number model with ?)

Answer: _____ markers

How many markers are left over?

_____ markers

SRB
63-64

---

2 Use doubling to help you solve.

$2 \times 7 =$ _____

$4 \times 7 =$ _____

_____ $= 2 \times 8$

_____ $= 4 \times 8$

SRB
49

---

3 Solve.

Estimate:

_____

$1,000 - 998 =$ _____

Unit

Think: Does my answer make sense?

SRB
103,
122-123

---

4 Show the time 8:48 on the clock.

What time will it be in

10 minutes? _____

SRB
18-19,
186

---

5 **Writing/Reasoning** Ava thinks counting up is the most efficient subtraction strategy for Problem 3. Do you agree? Explain.

Write another subtraction problem that you can solve efficiently by counting up. _____ − _____ = _____

How Much Money?

Math Message

1 Tildy and Harrison visit the dollar store. Tildy figures out how much she needs to pay by writing: $1 + 4 \times 2 = 10$. Tildy says that she needs $10. She asks Harrison to check her work, and he says she only needs $9.

Insert parentheses in the problems to show:

How Tildy got 10:

$1 + 4 \times 2 = 10$

How Harrison got 9:

$1 + 4 \times 2 = 9$

2 Tildy and Harrison could not agree about the correct amount. Tildy says, "I want to buy 1 package of paper for $1 and 4 fashion pens for $2 each."

Draw a picture to show what is happening in this problem.

3 Explain why Tildy changes her mind and now agrees with Harrison.

_____

_____

_____

Order of Operations

Use the order of operations to solve each number sentence below. Show your work. To check your work, use a calculator that follows the order of operations.

Rules for the Order of Operations

1. Do operations inside parentheses first.
   Follow rules 2 and 3 when computing inside parentheses.

2. Then multiply or divide, in order, from left to right.

3. Finally add or subtract, in order, from left to right.

① _____ = $11 - 3 \times 3$

② $15 \div 3 + 2 =$ _____

③ _____ = $20 \div 5 \times 2$

④ $6 + 4 \div 2 =$ _____

Try This

⑤ Circle the answer that makes the number sentence true.

$2 \times (4 + 3 \times 2) = ?$

a. 28

b. 20

c. 14

Explain to a partner why you picked your answer.

Math Boxes

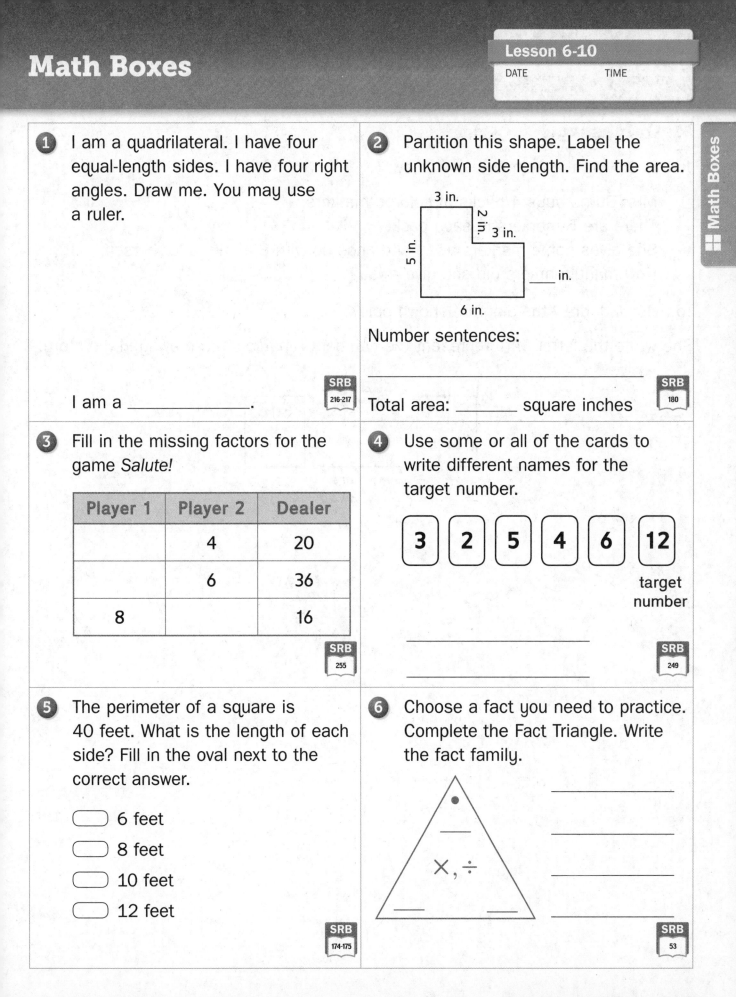

1 I am a quadrilateral. I have four equal-length sides. I have four right angles. Draw me. You may use a ruler.

I am a _____.

SRB 216-217

2 Partition this shape. Label the unknown side length. Find the area.

3 in.

2 in.   3 in.

5 in.

_____ in.

6 in.

Number sentences:

_____

Total area: _____ square inches

SRB 180

3 Fill in the missing factors for the game *Salute!*

Player 1	Player 2	Dealer
	4	20
	6	36
8		16

SRB 255

4 Use some or all of the cards to write different names for the target number.

3   2   5   4   6   12

target number

_____

_____

SRB 249

5 The perimeter of a square is 40 feet. What is the length of each side? Fill in the oval next to the correct answer.

( ) 6 feet

( ) 8 feet

( ) 10 feet

( ) 12 feet

SRB 174-175

6 Choose a fact you need to practice. Complete the Fact Triangle. Write the fact family.

×, ÷

_____

_____

_____

_____

SRB 53

two hundred thirteen   213

Representing a Number Story

Math Message

Judi is solving the number story below.

Mrs. Burns buys 4 packs of colored markers.
There are 5 markers in each pack.
She gives some markers away and ends up with 8 markers for herself.
How many markers did she give away?

To help, Judi drew the diagram shown below.

She wrote the letter *M* to represent the number of *markers* given away in the story.

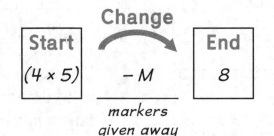

Number Models for Multistep Problems

For each problem:

- Write a number model. Use a letter for the unknown. You may draw a diagram.

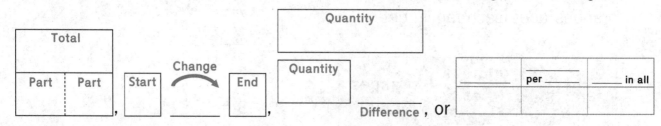

- Solve the problem. Show your work.

- Check that the answer makes your number model true. Write a summary number model.

1. Ronald bought 2 packs of crackers. There are 5 crackers in each pack. He ate some crackers. Now Ronald has 7 crackers. How many crackers did he eat?

   Letter and what it represents: _____ for _____

   _____
   (number model with letter)

   Answer: _____ crackers

   _____
   (number model with answer)

2. Leila bought 3 bags of fruit. Each bag has 4 oranges and 6 apples. How many pieces of fruit did Leila buy all together?

   Letter and what it represents: _____ for _____

   _____
   (number model with letter)

   Answer: _____ pieces of fruit

   _____
   (number model with answer)

Math Boxes

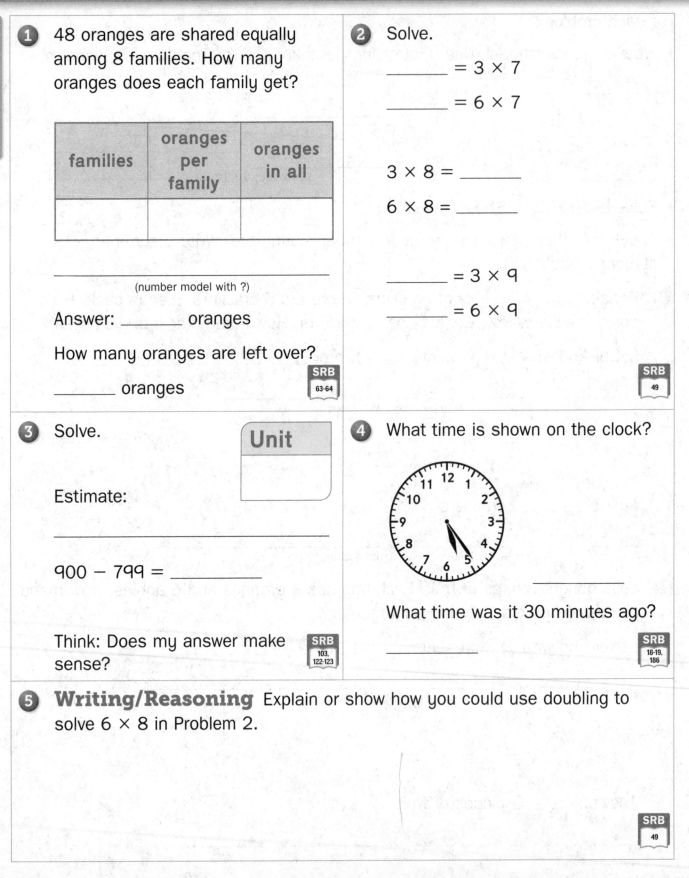

1 48 oranges are shared equally among 8 families. How many oranges does each family get?

families	oranges per family	oranges in all

_____
(number model with ?)

Answer: _____ oranges

How many oranges are left over?

_____ oranges

SRB
63-64

2 Solve.

_____ = 3 × 7

_____ = 6 × 7

3 × 8 = _____

6 × 8 = _____

_____ = 3 × 9

_____ = 6 × 9

SRB
49

3 Solve.

Unit

Estimate:

_____

900 − 799 = _____

Think: Does my answer make sense?

SRB
103,
122-123

4 What time is shown on the clock?

_____

What time was it 30 minutes ago?

_____

SRB
18-19,
186

5 **Writing/Reasoning** Explain or show how you could use doubling to solve 6 × 8 in Problem 2.

SRB
49

Math Boxes

Math Boxes
Preview for Unit 7

Math Boxes

1

1,000 mL   1,000 mL

500 mL   500 mL

Shade the empty beaker to show 1 liter of liquid.

How many liters of water are in both beakers together? ____ L

SRB
182

2 Label each part with a fraction.

Shade in three parts.

What fraction of the rectangle did you shade? _____

SRB
139

3 What fraction of each circle is shaded?

____   ____

Circle the larger fraction.

SRB
155

4 Kimani has an 18-inch long piece of construction paper. Circle ALL the ways he could cut it to make equal-length pieces.

A.   three 5-inch pieces

B.   two 9-inch pieces

C.   three 6-inch pieces

D.   six 3-inch pieces

SRB
40

5

0    $\frac{1}{2}$ 1   2   3   4   5   6

Inches (in.)

Mark the $\frac{1}{2}$ inch mark with a dot on the ruler.

Mark the $3\frac{1}{2}$ inch mark with an X on the ruler.

SRB
171

Using Liquid Volume Benchmarks

Look at the benchmark beakers to help estimate the volume of liquid your container can hold in liters or milliliters.

Then choose a beaker to measure how much liquid your container can hold.

1 My estimate: about _____   **2** Actual measure: about _____
(unit)   (unit)

3 Which benchmark beaker has a volume closest to your container's volume?

_____

How do you know? _____

_____

4 Which beaker did you choose to measure your container? _____

Why? _____

_____

5 Use a different beaker to measure the liquid volume of your container again.

What do you notice? _____

_____

Try This

6 Suppose your friend measured his container with the 100 mL beaker.

He emptied the beaker into his container 3 times. His container was still not full, so he emptied half of the 100 mL beaker into the container. Then it was full.

What is the liquid volume of his container? _____
(unit)

A Volume Puzzle

Allison has four jars.

A   B   C   D

1 Which jar is the tallest? _____

2 Which jar is the widest? _____

3 It takes 6 jar Ds to fill 1 jar A.

It takes $1\frac{1}{2}$ jar Ds to fill 1 jar B.

It takes 5 jar Ds to fill jar C.

Which jar has the most volume? Which has the least volume?
Explain your thinking.

_____

_____

_____

_____

University of Chicago

Solve these problems using the order of operations rules. Draw a line under the part of the number sentence that should be completed first.

Rules for the Order of Operations

1. If there are parentheses, do the operations inside the parentheses first. Follow rules 2 and 3 when computing inside parentheses.

2. Then multiply or divide, in order, from left to right.

3. Finally add or subtract, in order, from left to right.

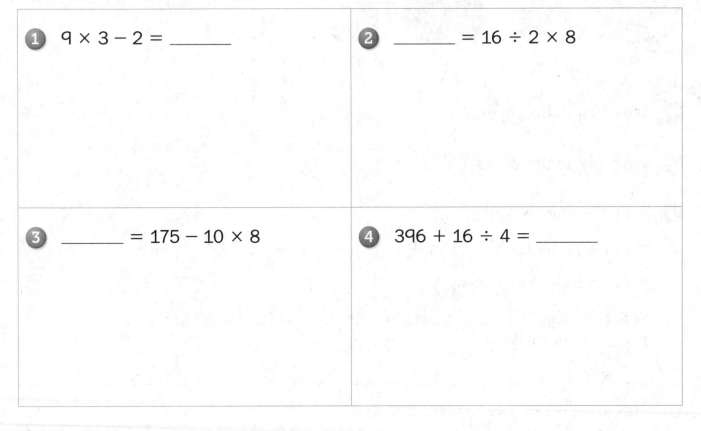

1 $9 \times 3 - 2 =$ _____

2 _____ $= 16 \div 2 \times 8$

3 _____ $= 175 - 10 \times 8$

4 $396 + 16 \div 4 =$ _____

5 Choose the number sentence from Problems 1–4 that matches this story. You may use a situation diagram.

Jeanne bought 3 bags of dog treats. There were 9 treats in each bag. Then Jeanne gave 2 treats to her dog. How many treats were left?

a. Number sentence: _____

b. Answer: _____
(unit)

Math Boxes

■■ Math Boxes

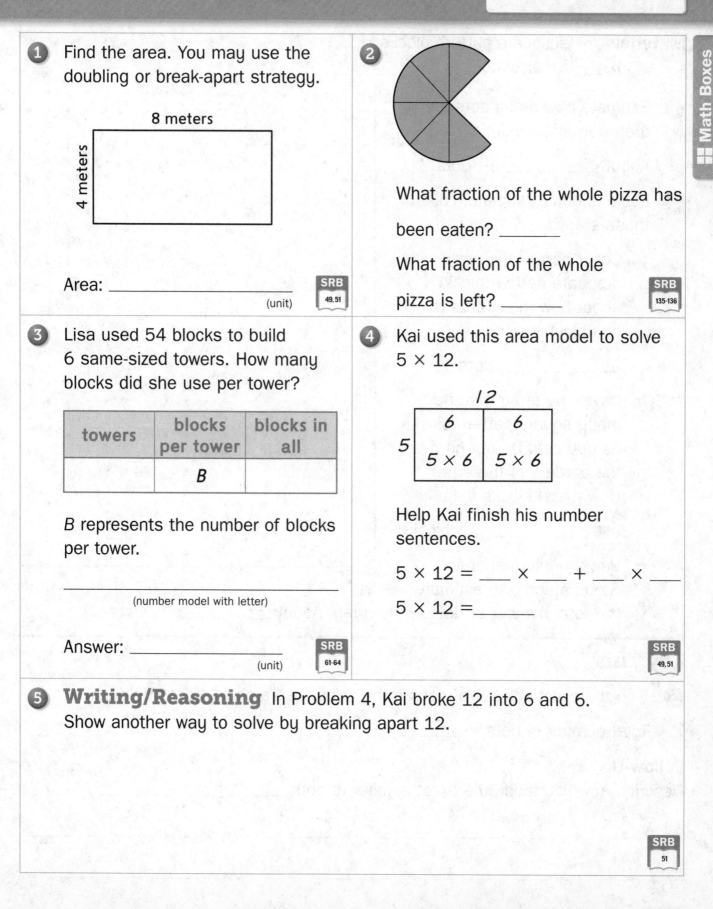

1 Find the area. You may use the doubling or break-apart strategy.

8 meters

4 meters

Area: _____
(unit)

SRB 49, 51

2

What fraction of the whole pizza has been eaten? _____

What fraction of the whole pizza is left? _____

SRB 135-136

3 Lisa used 54 blocks to build 6 same-sized towers. How many blocks did she use per tower?

towers	blocks per tower	blocks in all
	B	

B represents the number of blocks per tower.

_____
(number model with letter)

Answer: _____
(unit)

SRB 61-64

4 Kai used this area model to solve 5 × 12.

12

5   6   6

5 × 6   5 × 6

Help Kai finish his number sentences.

5 × 12 = ___ × ___ + ___ × ___

5 × 12 = _____

SRB 49, 51

5 **Writing/Reasoning** In Problem 4, Kai broke 12 into 6 and 6. Show another way to solve by breaking apart 12.

SRB 51

Exploration A: How Many Dots?

Materials   ☐ square pattern blocks
                  ☐ calculator

1 Estimate how many dots are in the array at the right.

About _____ dots

2 Make another estimate. Follow these steps.

   a. Cover part of the array with a square pattern block. About how many dots does one block cover?

     _____ dots

   b. Cover the array. Use as many square pattern blocks as you can. Do not go over the borders of the array. How many blocks did you use? _____ blocks

   c. Use the information in Steps a and b to estimate the total number of dots in the array. About _____ dots

Try This

3 Find the exact number of dots in the array. Use a calculator to help you.

Total number of dots = _____

Follow-Up

Describe how you found the exact number of dots. _____

_____

Math Boxes

① Solve.

$6 \div (2 + 1) =$ _____

$29 - (20 + 3) =$ _____

_____ $= (7 + 3) \times 2$

_____ $= (5 \times 5) - 6$

SRB
68

② Three fish tanks have 9 fish in each. One more tank has 5 fish. How many fish are there in all?

Circle the equation that fits the story. Then solve the story. *F* is the total number of fish.

$(3 \times 9) + 5 = F$

$(3 + 5) \times 9 = F$

_____ fish

SRB
68, 77

③ Erin helped her brother with homework. They worked for more than half an hour and finished at 6:50 P.M. Fill in the circles next to the times they could have started.

Ⓐ 6:45 P.M.

Ⓑ 6:15 P.M.

Ⓒ 6:30 P.M.

Ⓓ 6:05 P.M.

SRB
18-19

④ Solve.

$3 \times 5 =$ _____ $\times 3$

$3 \times 6 = 9 \times$ _____

$45 \div$ _____ $= 9$

_____ $\times 4 = 48 - 24$

SRB
61

⑤ Julie divided this rectilinear puppy pen into two smaller rectangles. Some of the sides are not labeled.

How long is the blue side? _____
(unit)

How do you know?

_____

_____

Area of the whole puppy pen: _____
(unit)

9 ft

6 ft

4 ft

4 ft

? ft

SRB
180-181

Math Boxes

Number Stories with Measures

Solve each problem. Show your work using drawings, number models, or words.
Remember to write the unit with each answer.

1 Kiana has 40 inches of green ribbon. She wants to cut it into nine 4-inch
pieces to make Earth Day award ribbons. Does Kiana have enough ribbon?

Answer: _____

How much ribbon, if any, would she have left over? _____
(unit)

2 The city zoo is home to a male African lion. After the lion arrived, he gained
about 8 kilograms of mass each month for 8 months. He now has a mass of
about 186 kilograms. What was the lion's starting mass?

Answer: about _____
(unit)

3 Lena has a doctor's appointment at 8:45 A.M. It takes her 25 minutes to
drive to her doctor's office. How many minutes early will Lena be if she leaves
at 8:00 A.M.?

Answer: about _____
(unit)

At what time must Lena leave to arrive at exactly 8:45 A.M.? _____

For Problems 4 and 5, use the pictures to help you solve.

4 How much more liquid is in Beaker B than in Beaker A?

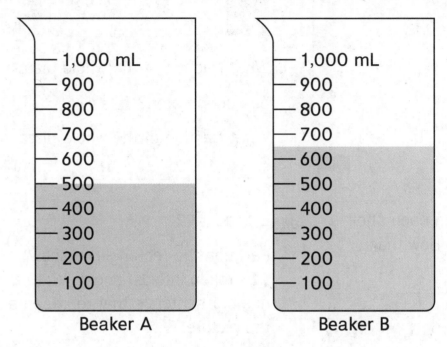

about _____
                              (unit)

5 Tomás wants to figure out the liquid volume of his glass. He fills a 50-milliliter beaker with water and empties it into his glass 4 times. His glass is now full of water. Then Tomás empties his glass into a 1-liter beaker. Shade the beaker to show how much water is in his glass.

Tomás's glass holds about _____
of water.   (unit)

Math Boxes

1 Find the area. You may use the doubling or break-apart strategy.

8 ft

6 ft

Area: _____
                    (unit)

SRB
49, 51

2

What fraction of the whole pizza

has been eaten? _____

What fraction of the whole pizza

is left? _____

SRB
135-136

3 There are 9 books on each shelf and 72 books in all. How many shelves are there?

shelves	books per shelf	books in all

_____ represents shelves
(letter)

_____
(number model with letter)

Answer: _____

SRB
61-64

4 6 × 11 = ?

Partition the rectangle to show 11 broken into 10 and 1. Write a number sentence that represents the picture.

11

6

6 × 11 = ___ × ___ + ___ × ___

6 × 11 = _____

SRB
51

5 **Writing/Reasoning** Use words, numbers, or drawings to show how you found the area in Problem 1.

SRB
49, 51

Comparing Fractions

Use your fraction strips to compare fractions.
Write <, >, or = between each pair of fractions
to make a true number sentence. For Problems 2–4,
draw your fraction strips to show each comparison.

Remember

= means *is equal to*

< means *is less than*

> means *is greater than*

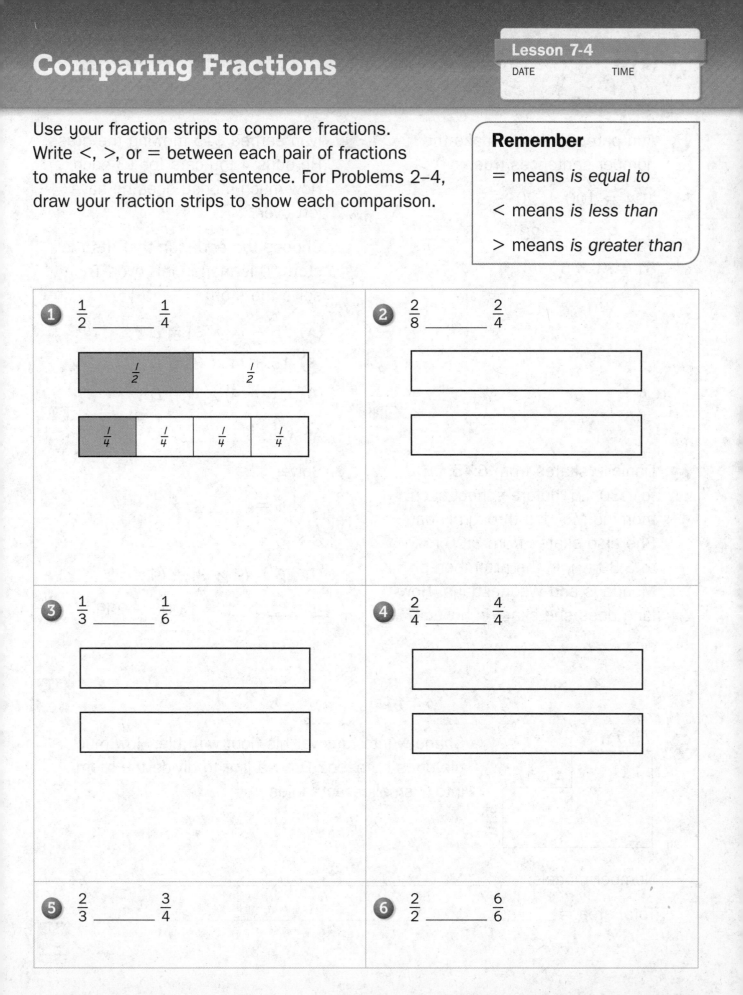

1 $\frac{1}{2}$ _____ $\frac{1}{4}$

| $\frac{1}{2}$ | $\frac{1}{2}$ |

| $\frac{1}{4}$ | $\frac{1}{4}$ | $\frac{1}{4}$ | $\frac{1}{4}$ |

2 $\frac{2}{8}$ _____ $\frac{2}{4}$

3 $\frac{1}{3}$ _____ $\frac{1}{6}$

4 $\frac{2}{4}$ _____ $\frac{4}{4}$

5 $\frac{2}{3}$ _____ $\frac{3}{4}$

6 $\frac{2}{2}$ _____ $\frac{6}{6}$

Math Boxes

1 Add parentheses to make the number sentences true.

$104 = 100 + 20 \div 5$

$4 \times 9 - 8 = 28$

$81 = 9 \times 5 + 4$

SRB
68

2 Ivan earned $15 mowing the grass. He buys 4 toy cars for $3 each. How much money does he have left over?

Choose the equation that fits the story. *D* is money left over. Then solve the story.

Ⓐ $15 - (4 \times 3) = D$

Ⓑ $15 + (4 \times 3) = D$

Ⓒ $(15 - 4) \times 3 = D$

Answer: $ _____

SRB
68, 77

3 Danielle skates from 6:45 A.M. to 7:30 A.M. before school each morning Monday through Friday. She also skates from 3:00 P.M. to 3:30 P.M. in the afternoon on Mondays and Wednesdays. How long does she skate in a week?

_____

SRB
18-19,
187-188

4 Solve.

$7 \times 6 = $ _____ $\times 7$

$8 \times 6 = 12 \times$ _____

$9 \times 6 = (9 \times 3) + (9 \times$ _____$)$

_____ $\times 1 = 1,000 - 999$

SRB
61

5

7 ft

3 ft

6 ft

5 ft

Shane wants to cover his floor with tile. How much tile does he need? Draw a line to divide the room into 2 smaller rectangles.

Number models: _____

Total area = _____

(unit)

SRB
180-181

Fraction Number-Line Poster

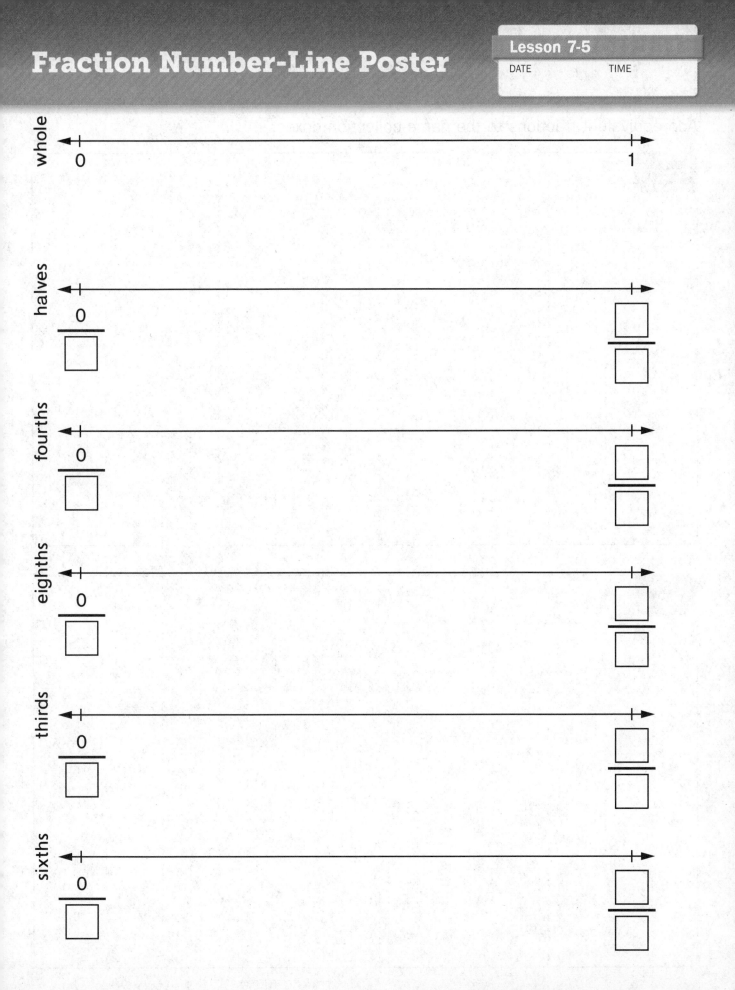

Fraction Name-Collection Boxes

Add equivalent fractions to the name-collection boxes.

$\frac{2}{3}$

$\frac{3}{4}$

Math Boxes

① Complete the Fact Triangle. Write the fact family.

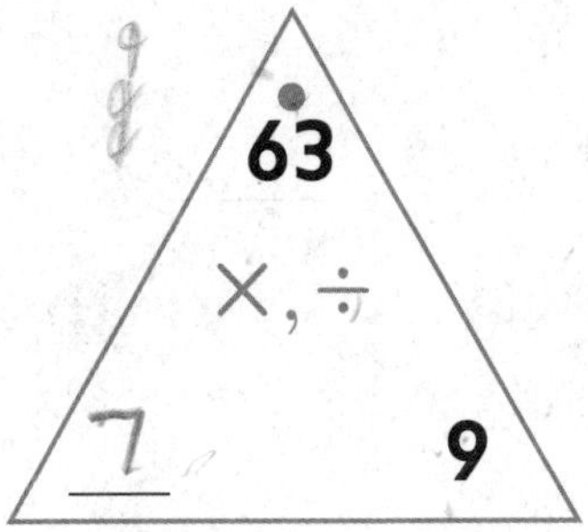

$$63 \div 9 = 7$$
$$63 \div 7 = 9$$
$$9 \times 7 = 63$$
$$7 \times 9 = 63$$

Triangle with 63 at top, ×, ÷ in middle, 7 and 9 at bottom corners.

SRB 53

② Amanda has 16 liters of water. How many 2-liter bottles can she fill?

$$2 \times 8 = 16$$

Answer: __8__ bottles

SRB 182

③ Write or draw three equivalent fractions.

$\frac{1}{2}$ $\frac{2}{4}$

SRB 152-153

④ Use order of operations rules. Underline the part that should be completed first. Then solve.

$10 - 2 \times 2 =$ _____

$20 + 10 \div 5 =$ _____

SRB 69

⑤ Trish has 3 bags of apples. Each bag has 4 apples. Trish gives 7 apples away. How many apples does she have left?

Answer: _____ apples

Circle the equation that fits the story. A is the number of apples left.

$(3 \times 4) - 7 = A$

$3 \times (7 - 4) = A$

SRB 68, 77

⑥ Solve.

Estimate: _____

$562 - 177 = ?$

Ⓐ 415

Ⓑ 485

Ⓒ 385

Ⓓ 215

Unit

Think: Does my answer make sense?

SRB 103, 119-123

Fractions on Number Lines

Number Line A

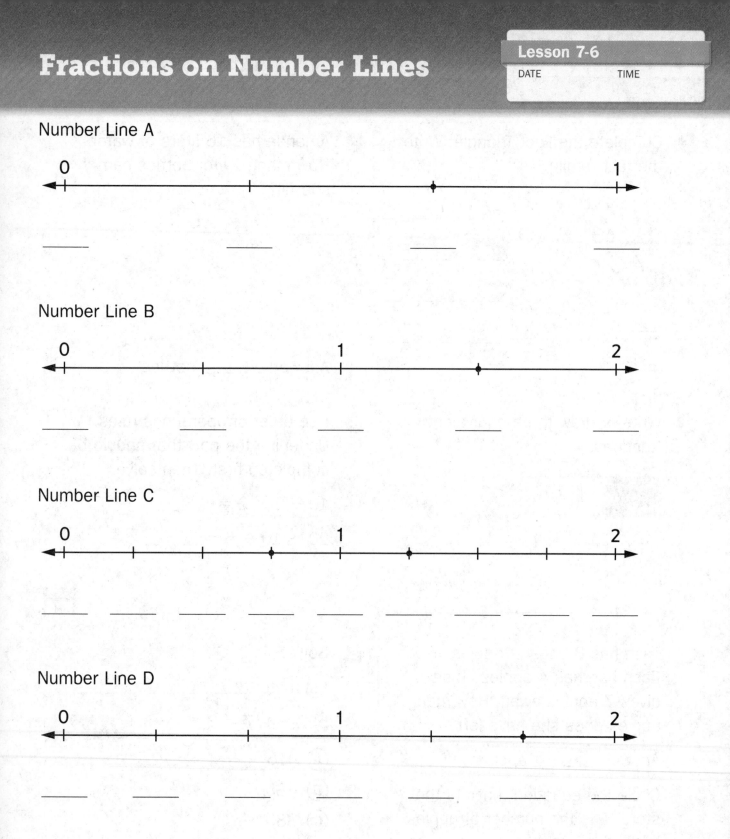

Number Line B

Number Line C

Number Line D

Naming Fractions on Number Lines

Write the missing fractions on the number lines. Then write the fraction that names each point.

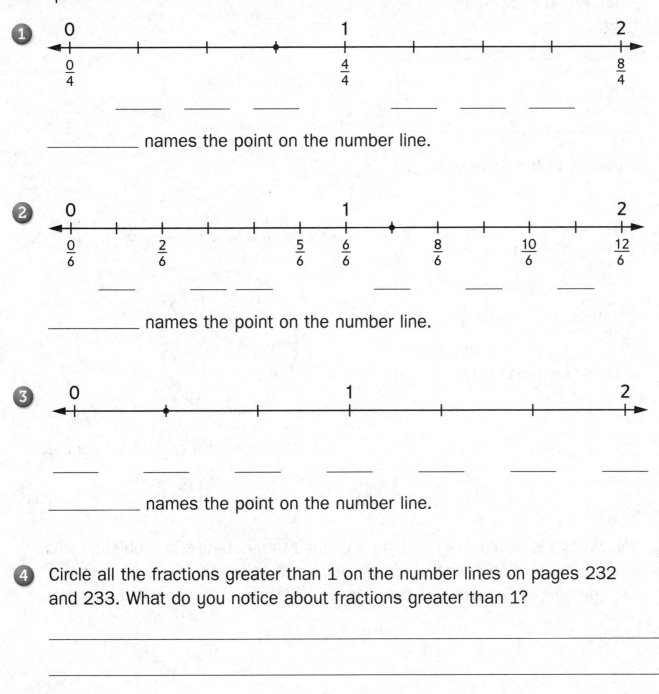

1.

$\frac{0}{4}$ $\frac{4}{4}$ $\frac{8}{4}$

_____ names the point on the number line.

2.

$\frac{0}{6}$ $\frac{2}{6}$ $\frac{5}{6}$ $\frac{6}{6}$ $\frac{8}{6}$ $\frac{10}{6}$ $\frac{12}{6}$

_____ names the point on the number line.

3.

_____ names the point on the number line.

4. Circle all the fractions greater than 1 on the number lines on pages 232 and 233. What do you notice about fractions greater than 1?

_____

_____

_____

Math Boxes

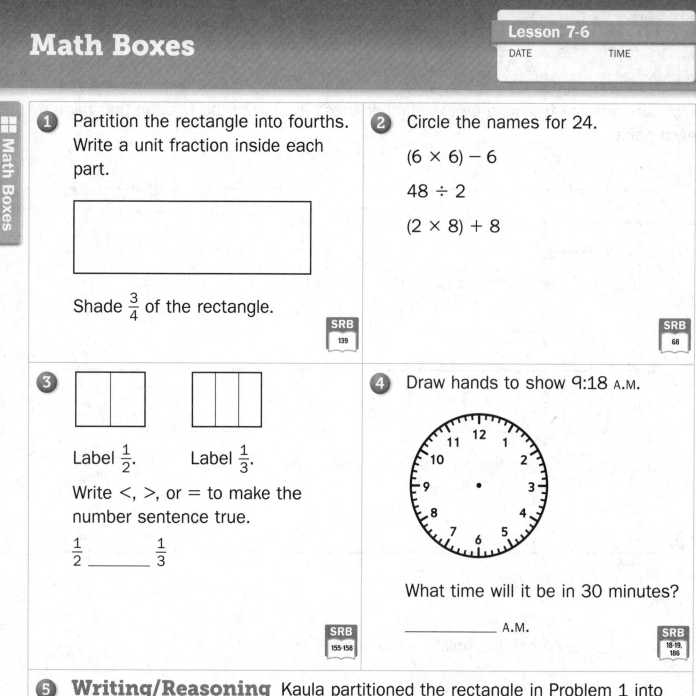

1 Partition the rectangle into fourths. Write a unit fraction inside each part.

Shade $\frac{3}{4}$ of the rectangle.

SRB
139

2 Circle the names for 24.

$(6 \times 6) - 6$

$48 \div 2$

$(2 \times 8) + 8$

SRB
68

3

Label $\frac{1}{2}$.   Label $\frac{1}{3}$.

Write <, >, or = to make the number sentence true.

$\frac{1}{2}$ _____ $\frac{1}{3}$

SRB
155-158

4 Draw hands to show 9:18 A.M.

What time will it be in 30 minutes?

_____ A.M.

SRB
18-19,
186

5 **Writing/Reasoning** Kayla partitioned the rectangle in Problem 1 into 8 equal parts. She shaded 6 of the parts and said that $\frac{6}{8}$ is equivalent to $\frac{3}{4}$. Do you agree? Explain. You may use your drawing in Problem 1 to help you.

SRB
150-153

Fractions Greater Than and Less Than $\frac{1}{2}$

Use the Fraction Number-Line Poster to help you write fractions that are less than, equal to, and greater than $\frac{1}{2}$.

1 Fill in the table.

Fraction Benchmarks		
Fractions Less Than $\frac{1}{2}$	**Fractions Equal to $\frac{1}{2}$**	**Fractions Greater Than $\frac{1}{2}$**

2 Choose two fractions that are less than $\frac{1}{2}$.

_____ _____

Of these two fractions, which one is closer to 0?

How do you know? _____

_____

Write a number sentence that compares your two fractions. Use <, >, or =.

_____ ☐ _____

> **Remember**
>
> = means *is equal to*
>
> < means *is less than*
>
> > means *is greater than*

3 Choose two fractions that are greater than $\frac{1}{2}$. _____ _____

Of these two fractions, which one is closer to 1? How do you know?

_____

Write a number sentence that compares your two fractions. Use <, >, or =.

_____ ☐ _____

Exploring Shape and Volume

Use the information and pictures below to help you answer the question.

Gail and Keron fill container A with water.

They pour all of the water from container A into container B. Now container B is full.

Then they pour all of the water from container B into container C. Now container C is full.

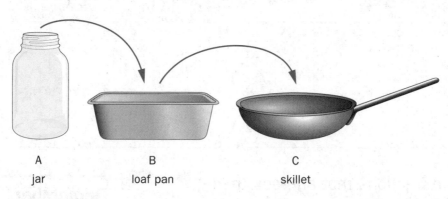

A
jar

B
loaf pan

C
skillet

Gail says the liquid volume of container C is *less than* the liquid volume of container A. She says that container C holds less water because it is wide and short, and container A holds more because it is taller.

Keron says all three of their containers have about the same volume.

Do you agree with Gail or Keron? Explain your thinking.

Math Boxes

1 Fill in the Fact Triangle. Write the fact family.

_____

_____

$\times, \div$

_____

_____

SRB
53

2 Justin's grandma poured 30 liters of water into 5-liter jugs. How many 5-liter jugs did she fill?

Answer: _____ jugs

SRB
182

3 Fill in the circles next to the names that belong in this name-collection box.

$\frac{2}{2}$

(A) 1   (B) $\frac{3}{3}$

(C) $\frac{1}{8}$ (D) $\frac{8}{8}$

SRB
152-154

4 Use order of operations rules. Underline the part that should be completed first. Then solve.

_____ $= 30 - 15 \div 3$

$8 \times 6 - 6 =$ _____

SRB
69

5 Charlie now has 20 batteries. He started with 6 packs of 10 batteries. How many batteries did Charlie use?

Answer: _____ batteries

Circle the equation that fits the story. *B* is the number of batteries Charlie used.

$20 = (6 \times 10) - B$

$B = (6 \times 10) + 20$

SRB
68, 77

6 Make an estimate. Solve.

Estimate: _____

$479 + 356 =$ _____

Unit

Think: Does my answer make sense?

SRB
103,
116-118

Comparing Fractions to $\frac{1}{2}$

Steve is sorting the number sides of his fraction cards into two groups: fractions greater than $\frac{1}{2}$ and fractions less than $\frac{1}{2}$.

He made a mistake.

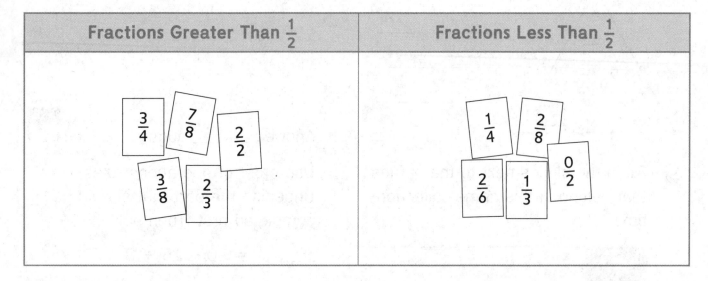

Fractions Greater Than $\frac{1}{2}$	Fractions Less Than $\frac{1}{2}$
$\frac{3}{4}$ $\frac{7}{8}$ $\frac{2}{2}$ $\frac{3}{8}$ $\frac{2}{3}$	$\frac{1}{4}$ $\frac{2}{8}$ $\frac{2}{6}$ $\frac{1}{3}$ $\frac{0}{2}$

① What mistake did Steve make? Use your Fraction Number-Line Poster, fraction strips, fraction circles, or fraction cards to help you. Mark an X on the card that Steve put in the wrong group. Show or explain how you decided.

② With a partner, write a rule that Steve can use to check whether a fraction is greater than $\frac{1}{2}$ or less than $\frac{1}{2}$.

_____

_____

Math Boxes

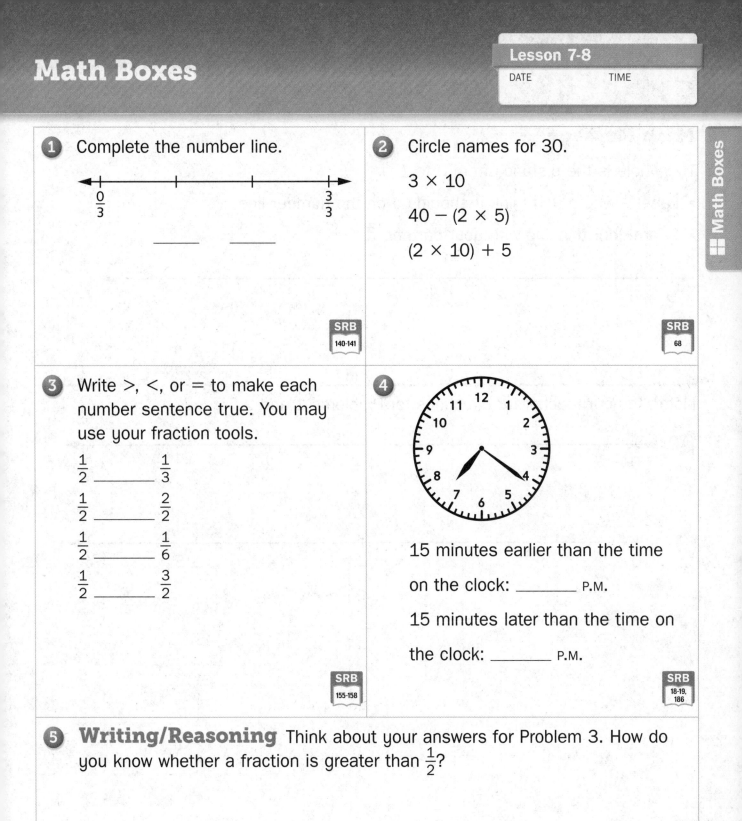

1 Complete the number line.

$$\frac{0}{3} \qquad\qquad\qquad\qquad \frac{3}{3}$$

_____   _____

SRB
140-141

2 Circle names for 30.

3 × 10

40 − (2 × 5)

(2 × 10) + 5

SRB
68

3 Write >, <, or = to make each number sentence true. You may use your fraction tools.

$\frac{1}{2}$ _____ $\frac{1}{3}$

$\frac{1}{2}$ _____ $\frac{2}{2}$

$\frac{1}{2}$ _____ $\frac{1}{6}$

$\frac{1}{2}$ _____ $\frac{3}{2}$

SRB
155-158

4

15 minutes earlier than the time

on the clock: _____ P.M.

15 minutes later than the time on

the clock: _____ P.M.

SRB
18-19,
186

5 **Writing/Reasoning** Think about your answers for Problem 3. How do you know whether a fraction is greater than $\frac{1}{2}$?

SRB
156

Math Message

The whole is the distance from 0 to 1.

- Label $\frac{1}{2}$ where you think it should go on the number line.
- Share your thinking with your partner.

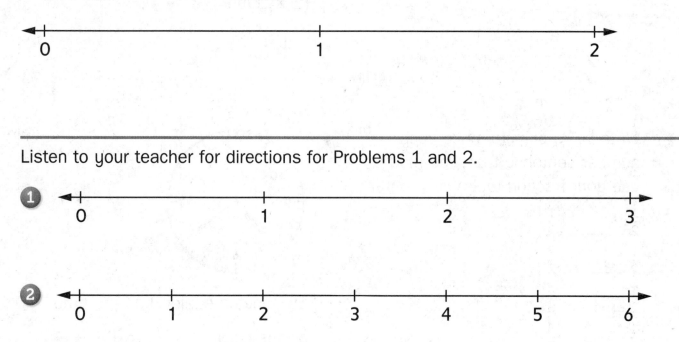

Listen to your teacher for directions for Problems 1 and 2.

Locating Fractions on Number Lines

The whole on the number lines below is the distance between 0 and 1. Partition each number line to locate and then label the given fraction. *Hint:* On some number lines, there is more than 1 whole.

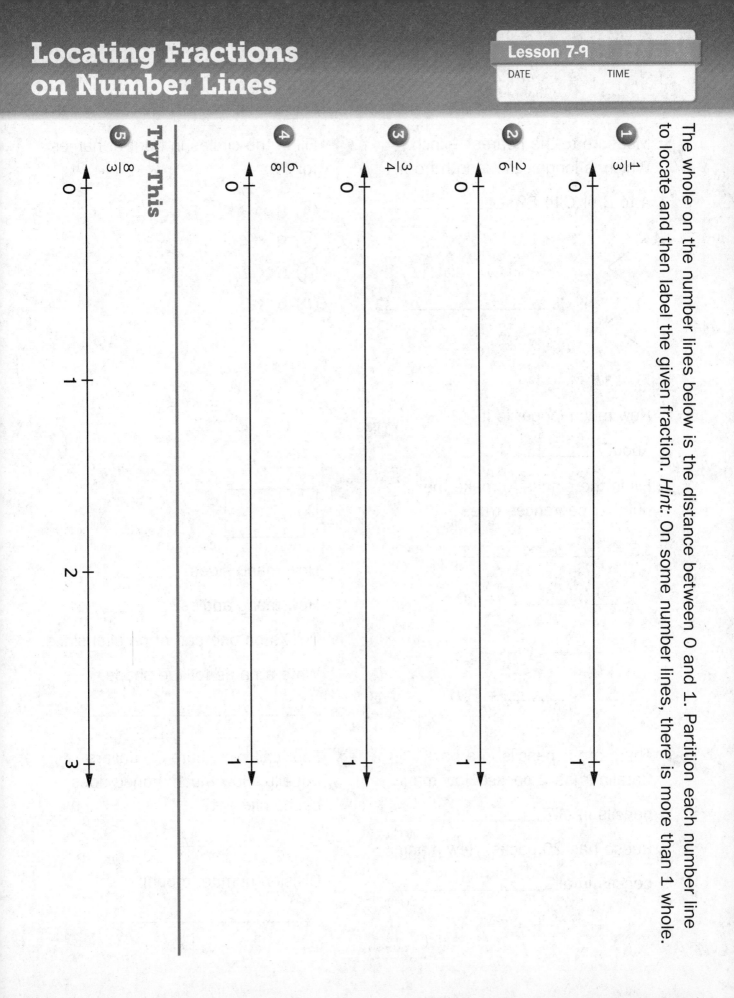

① $\frac{1}{3}$

0 ———————————————— 1

② $\frac{2}{6}$

0 ———————————————— 1

③ $\frac{3}{4}$

0 ———————————————— 1

④ $\frac{6}{8}$

0 ———————————————— 1

Try This

⑤ $\frac{8}{3}$

0 ———— 1 ———— 2 ———— 3

Math Boxes
Preview for Unit 8

① Measure to the nearest $\frac{1}{2}$ inch. Which is longer, the length from

A to B or C to D? _____

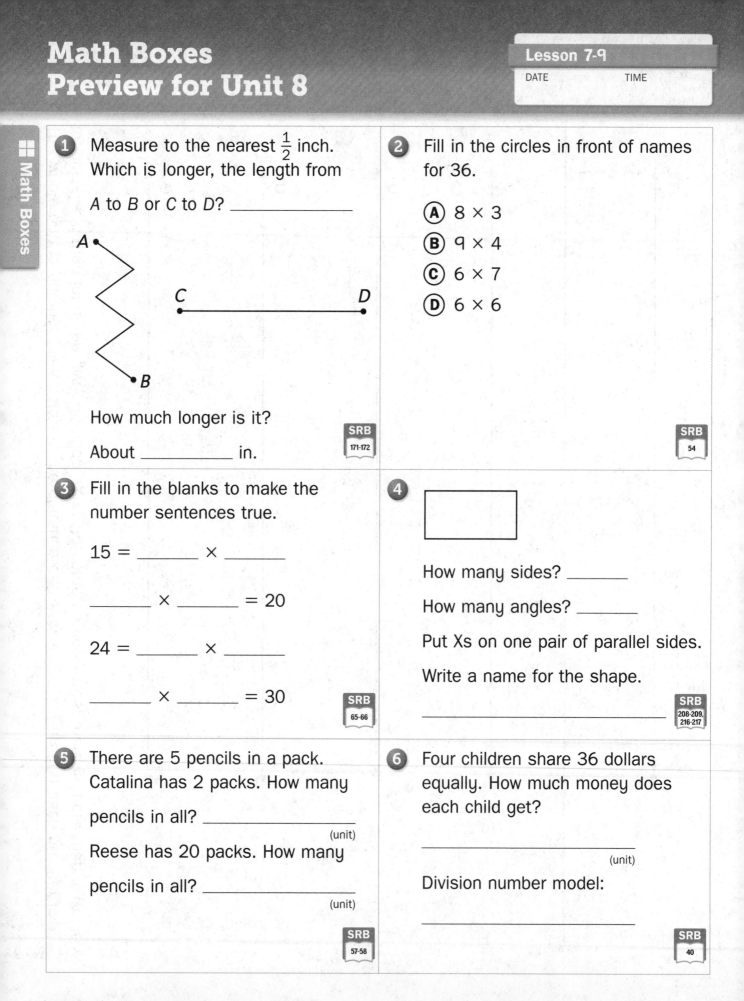

How much longer is it?

About _____ in.

SRB
171-172

② Fill in the circles in front of names for 36.

Ⓐ 8 × 3

Ⓑ 9 × 4

Ⓒ 6 × 7

Ⓓ 6 × 6

SRB
54

③ Fill in the blanks to make the number sentences true.

15 = _____ × _____

_____ × _____ = 20

24 = _____ × _____

_____ × _____ = 30

SRB
65-66

④

How many sides? _____

How many angles? _____

Put Xs on one pair of parallel sides.

Write a name for the shape.

_____

SRB
208-209,
216-217

⑤ There are 5 pencils in a pack. Catalina has 2 packs. How many

pencils in all? _____
                              (unit)

Reese has 20 packs. How many

pencils in all? _____
                              (unit)

SRB
57-58

⑥ Four children share 36 dollars equally. How much money does each child get?

_____
                              (unit)

Division number model:

_____

SRB
40

242   two hundred forty-two

Modeling Equivalent Fractions

Name a fraction that is equivalent to $\frac{2}{3}$. You may use your fraction tools to help.

$\frac{2}{3}$ = _____

Show that your number sentence is true with number lines, fraction strips, and fraction circle pieces. Sketch pictures below to show what you did.

Number Lines

0   1

0   1

Fraction Strips

Fraction Circle Pieces

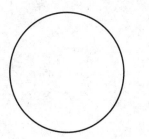 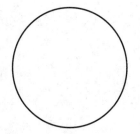

Justifying Fraction Comparisons

Use fraction circles, fraction strips, and fraction number lines to solve. Use at least two different fraction tools to show that each number sentence is true. Sketch pictures to show what you did.

1. Find two equivalent fractions to make this number sentence true: $\dfrac{\Box}{\Box} = \dfrac{\Box}{\Box}$

   My sketches:

2. Find two fractions to make this number sentence true: $\dfrac{\Box}{\Box} > \dfrac{\Box}{\Box}$

   My sketches:

Math Boxes

① The triangle started at 0. Write a fraction to show how far it moved.

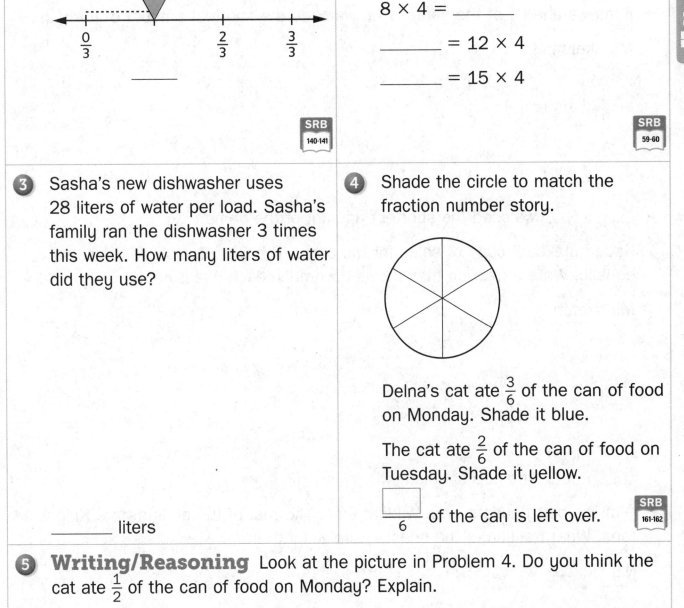

$\dfrac{0}{3}$ $\dfrac{2}{3}$ $\dfrac{3}{3}$

_____

SRB
140-141

② Solve.

$7 \times 4 =$ _____

$8 \times 4 =$ _____

_____ $= 12 \times 4$

_____ $= 15 \times 4$

SRB
59-60

③ Sasha's new dishwasher uses 28 liters of water per load. Sasha's family ran the dishwasher 3 times this week. How many liters of water did they use?

_____ liters

④ Shade the circle to match the fraction number story.

Delna's cat ate $\dfrac{3}{6}$ of the can of food on Monday. Shade it blue.

The cat ate $\dfrac{2}{6}$ of the can of food on Tuesday. Shade it yellow.

$\dfrac{\boxed{}}{6}$ of the can is left over.

SRB
161-162

⑤ **Writing/Reasoning** Look at the picture in Problem 4. Do you think the cat ate $\dfrac{1}{2}$ of the can of food on Monday? Explain.

SRB
150-152

Football Party Fractions

Use fraction circles, fraction strips, number lines, or pictures to help solve the number stories. Make sketches to show how you solved.

1. Tania watched the whole football game, Micah watched $\frac{1}{4}$ of the game, and Alma watched $\frac{3}{4}$ of the game. Who watched the shortest amount of the game?

   My sketch:

   _____ watched the shortest amount of the game.

2. Kaden makes 2 cups of salsa for the party. The 6 guests share the salsa equally. Write a fraction that shows how much each guest eats.

   My sketch:

   _____ of a cup of salsa

3. $\frac{1}{6}$ of the party guests were Wildcat fans. The rest of the guests were Knight fans. What fraction of the guests were Knight fans?

   My sketch:

   _____ of the guests

Fractions on Number Lines

The whole on a number line is the distance from 0 to 1. Each number line below shows more than 1 whole. Label the wholes. Then partition to locate and label the given fraction.

1. $\frac{1}{2}$

2. $\frac{2}{4}$

Write a fraction below each triangle to show the distance from 0.

3.

4.

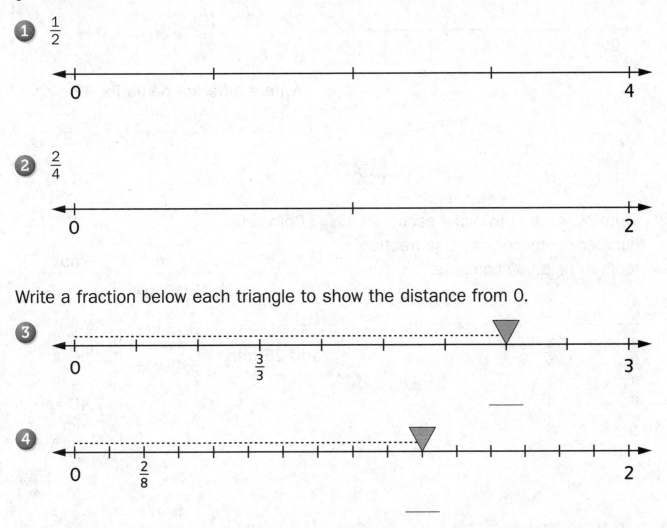

5. Label at least two more fractions on the number lines in Problems 3 and 4.

Math Boxes

1 Partition to show fourths.

Label $\frac{3}{4}$.

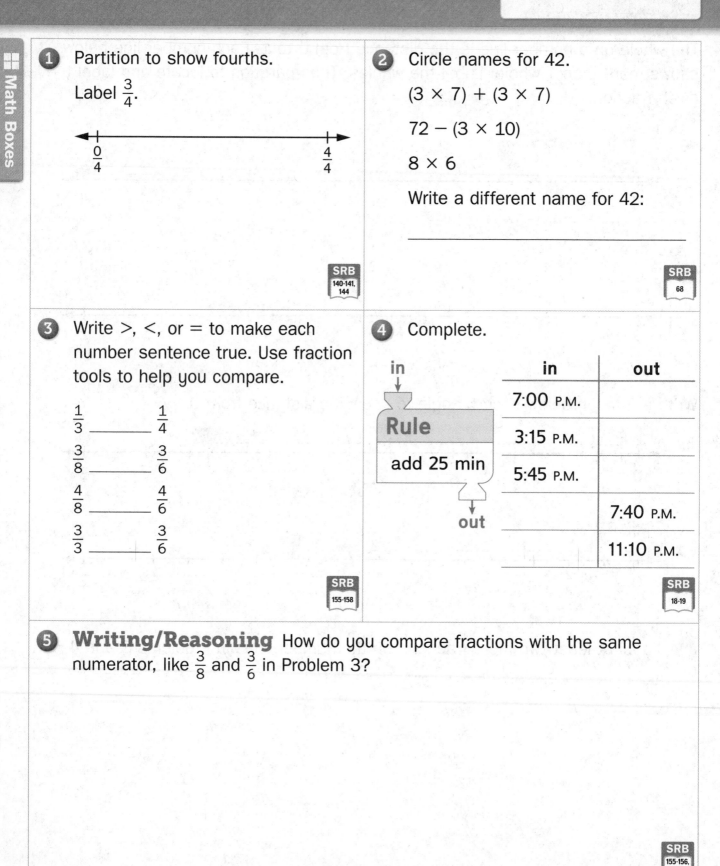

$\frac{0}{4}$ $\frac{4}{4}$

SRB
140-141,
144

2 Circle names for 42.

$(3 \times 7) + (3 \times 7)$

$72 - (3 \times 10)$

8×6

Write a different name for 42:

_____

SRB
68

3 Write >, <, or = to make each number sentence true. Use fraction tools to help you compare.

$\frac{1}{3}$ _____ $\frac{1}{4}$

$\frac{3}{8}$ _____ $\frac{3}{6}$

$\frac{4}{8}$ _____ $\frac{4}{6}$

$\frac{3}{3}$ _____ $\frac{3}{6}$

SRB
155-158

4 Complete.

in

Rule

add 25 min

out

in	out
7:00 P.M.	
3:15 P.M.	
5:45 P.M.	
	7:40 P.M.
	11:10 P.M.

SRB
18-19

5 **Writing/Reasoning** How do you compare fractions with the same numerator, like $\frac{3}{8}$ and $\frac{3}{6}$ in Problem 3?

SRB
155-156,
158

Solving Number Stories Using Collections

Solve. You may use counters and draw pictures to help.

1. Two people share a collection of 8 pennies equally. How many pennies does each person get? _____ pennies

   What fraction of the pennies in the collection does each person get?

   _____ of the pennies

2. Five people share a collection of 15 pennies equally. How many pennies does each person get? _____ pennies

   What fraction of the pennies in the collection does each person get?

   _____ of the pennies

3. There are 12 eggs in the carton. After the carton is dropped, 7 of the eggs are cracked.

   What fraction of the eggs in the carton did not crack? _____ of the eggs

4. Julie and Amanda each have 8 pencils. $\frac{3}{8}$ of Julie's pencils are blue. $\frac{5}{8}$ of Amanda's pencils are blue. Who has more blue pencils?

   Draw a picture to show your thinking.

   _____ has more blue pencils.

Math Boxes

① The triangle started at 0. Write a fraction to show how far it moved.

$\frac{0}{4}$ $\frac{4}{4}$

_____

SRB
140-141

② Solve.

$8 \times 6 =$ _____

$9 \times 6 =$ _____

_____ $= 12 \times 6$

_____ $= 15 \times 6$

SRB
59-60

③ A 10-minute shower uses 190 liters of water. A 5-minute shower uses 95 liters of water.

How many liters of water can you save by taking a 5-minute shower instead of a 10-minute shower?

_____ liters

SRB
122-123

④ Complete the fraction number story.

Margaret ate $\dfrac{\boxed{}}{8}$ of the pizza.

Justin ate $\dfrac{\boxed{}}{8}$ of the pizza.

Connor ate $\dfrac{\boxed{}}{8}$ of the pizza.

$\dfrac{\boxed{}}{8}$ of the pizza was left over.

SRB
161-162

⑤ **Writing/Reasoning** Explain how you could use either the doubling or break-apart strategy to solve 15×6 in Problem 2.

SRB
49, 51

Math Boxes
Preview for Unit 8

① Label $\frac{1}{8}$ and $\frac{1}{4}$ on the number line. You may use your fraction tools.

0          $\frac{1}{2}$ 1

SRB
140, 145

② Fill in the circle next to each name for 40.

(A) 4 × 10

(B) 8 × 5

(C) 2 × 20

(D) 8 × 4

SRB
54

③ Fill in the blanks to make the number sentences true.

_____ × _____ = 36

_____ × _____ = 45

50 = _____ × _____

_____ × _____ = 64

SRB
65-66

④

How many sides? _____

How many angles? _____

Put Xs on one pair of parallel sides.

Write a name for the shape.

_____

SRB
208-209, 216-217

⑤ There are 8 tissue boxes in a pack. Clara takes 2 packs to school. How many tissue boxes in all?

_____
(unit)

Clara's class now has 20 packs. How many tissue boxes in all?

_____
(unit)

SRB
57-58

⑥ Dontrell shares 24 dollars equally with his brother. How much money do they each get?

_____
(unit)

Division number model:

_____

SRB
40

Math Boxes

two hundred fifty-one   251

Measuring to the Nearest $\frac{1}{4}$ Inch

Use Ruler D to measure the line segments below to the nearest $\frac{1}{4}$ inch.

1 _____

about _____ in.

2 _____

about _____ in.

3 _____

about _____ in.

Fractions on a Number Line

① Think about where each of the fractions below belong on the number line. Then write one of the fractions in each box for *A*, *B*, *C*, and *D* on the number line.

$\frac{1}{2}$ $\frac{1}{3}$ $\frac{1}{4}$ $\frac{3}{4}$

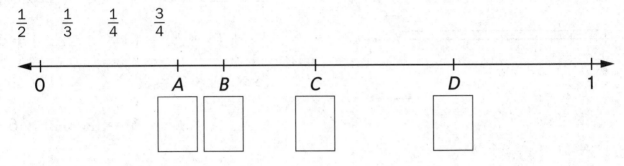

Explain how you figured out the location of $\frac{3}{4}$ on the number line.

_____

_____

What is another fraction name for the point you labeled $\frac{1}{2}$? _____

② Look at the number line below. Label the tick mark with a fraction that could name that point. (*Hint:* You may partition the number line to help you identify the point.)

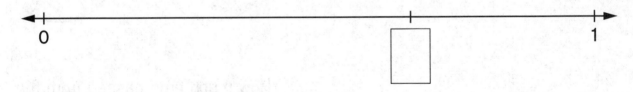

Try This

③ Label $\frac{3}{6}$ on the number line in Problem 2.

Explain how you know where $\frac{3}{6}$ belongs.

_____

_____

Math Boxes

1 What fraction of the distance from 0 to 1 did the triangle move?

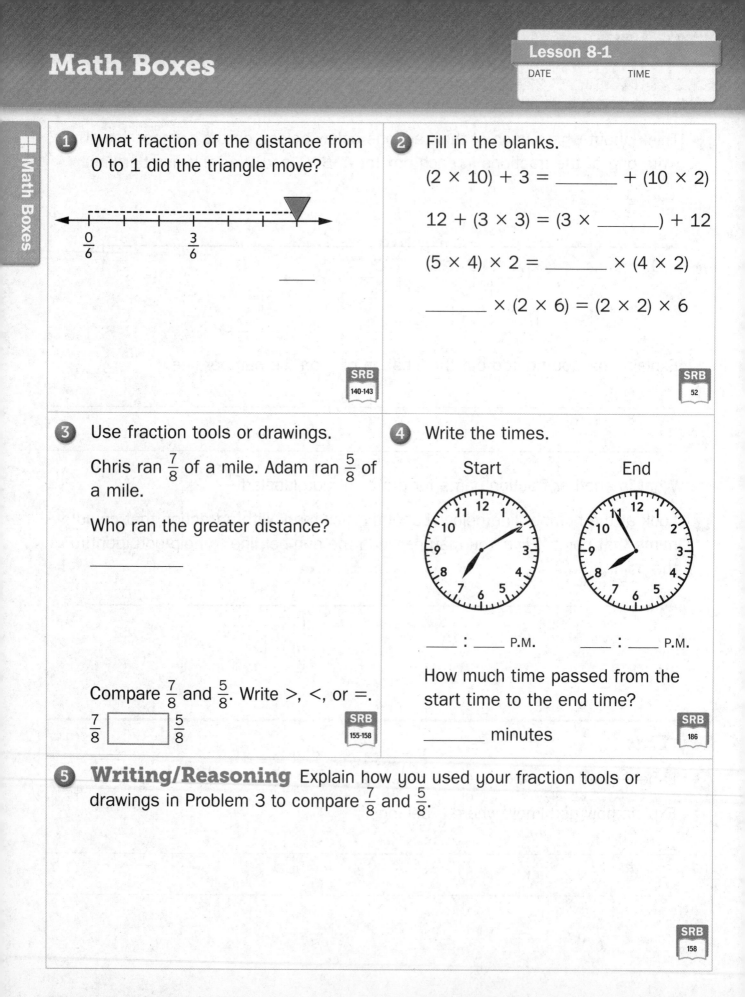

$\frac{0}{6}$ $\frac{3}{6}$

_____

SRB
140-143

2 Fill in the blanks.

$(2 \times 10) + 3 =$ _____ $+ (10 \times 2)$

$12 + (3 \times 3) = (3 \times$ _____$) + 12$

$(5 \times 4) \times 2 =$ _____ $\times (4 \times 2)$

_____ $\times (2 \times 6) = (2 \times 2) \times 6$

SRB
52

3 Use fraction tools or drawings.

Chris ran $\frac{7}{8}$ of a mile. Adam ran $\frac{5}{8}$ of a mile.

Who ran the greater distance?

_____

Compare $\frac{7}{8}$ and $\frac{5}{8}$. Write >, <, or =.

$\frac{7}{8}$ ☐ $\frac{5}{8}$

SRB
155-158

4 Write the times.

Start   End

____ : ____ P.M.   ____ : ____ P.M.

How much time passed from the start time to the end time?

_____ minutes

SRB
186

5 **Writing/Reasoning** Explain how you used your fraction tools or drawings in Problem 3 to compare $\frac{7}{8}$ and $\frac{5}{8}$.

SRB
158

Extended Multiplication Facts

Show each set of problems using base-10 blocks. Use ∎ | ☐ to show what
you did. Then write the products.

Set 1:

2 × 3 = _____   2 × 30 = _____   2 × 300 = _____

Set 2:

3 × 5 = _____   3 × 50 = _____   3 × 500 = _____

What do you notice in the products in each set?

_____

_____

Extended Multiplication and Division Facts

Identify a basic fact that you can use to help you solve the following extended facts. Then solve.

 a. $5 \times 70 = ?$

Basic fact that I can use to help: _____

$5 \times 70 =$ _____

b. Explain how you solved 5×70.

_____

_____

 $8 \times 30 = ?$

Basic fact that I can use to help: _____

$8 \times 30 =$ _____

3 $6 \times 80 = ?$

Basic fact that I can use to help: _____

$6 \times 80 =$ _____

 $320 \div 4 = ?$

Basic fact that I can use to help: _____

$320 \div 4 =$ _____

5 $540 \div 6 = ?$

Basic fact that I can use to help: _____

$540 \div 6 =$ _____

A New Classroom Bookcase

Your teacher is buying a new bookcase for the classroom. There are three shelves that can hold books of different heights.

 Choose three books from your classroom and measure their heights to the nearest $\frac{1}{4}$ inch.

Height of book 1: about _____ inches

Height of book 2: about _____ inches

Height of book 3: about _____ inches

2 Figure out the *shortest* shelf on which book 1 could fit. Write book 1 next to that shelf. Do the same thing for book 2 and book 3.

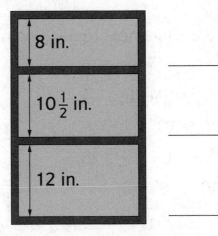

Do you have a book that is too tall to fit on the shelf? _____

If one of your books is too tall to fit, how many inches tall would a shelf need

to be for the book to fit? _____

3 Plot your book measurements on the class line plot.

Math Boxes

Math Boxes

1 Four friends equally share these books. Shade the number of books in one equal share.

What fraction of books is shaded?

_____

SRB
40, 146

2 Estimate the perimeter of a basketball court and then find the exact perimeter.

Estimate:

_____

94 feet

50 feet

Perimeter: _____
(unit)

SRB
116-118, 174

3 Write >, <, or = to make each number sentence true. Use the Fraction Number-Line Poster to help.

$\frac{3}{4}$ ___ 1 $\frac{1}{4}$ ___ 1

$\frac{5}{4}$ ___ 1 $\frac{2}{4}$ ___ 1

$\frac{4}{4}$ ___ 1 $\frac{12}{4}$ ___ 1

SRB
145, 155-158

4 Fill in the circle next to each name for 200.

Ⓐ (4 × 10) + 160

Ⓑ 100 + (9 × 10)

Ⓒ 500 − (175 + 125)

Ⓓ 240 − (8 × 5)

SRB
57

5 Fill in the rule and the missing numbers.

in

Rule

out

in	out
4	16
	8
6	
8	
10	40

SRB
74

6 Write a fraction to make the number sentence true.

$\frac{1}{3}$ = _____

Draw a picture to show that your number sentence is true.

SRB
152

Factors of Products

For each problem, fill in the blanks with different factor pairs that make the number sentences true.

1. 8 = _____ × _____   8 = _____ × _____

2. 18 = _____ × _____   18 = _____ × _____

3. 27 = _____ × _____   27 = _____ × _____

4. 36 = _____ × _____   36 = _____ × _____

5. 45 = _____ × _____   45 = _____ × _____

6. The Kim family is serving dinner for 24 people. Mrs. Kim could have 1 table with 24 people or 2 tables with 12 people each.

   What are some other ways Mrs. Kim could seat 24 people in equal groups at different numbers of tables? (*Hint:* Think of factor pairs for 24.) Use pictures, words, or numbers to show your table arrangements.

Try This

7. 200 = _____ × _____   200 = _____ × _____

8. 720 = _____ × _____   720 = _____ × _____

Math Boxes

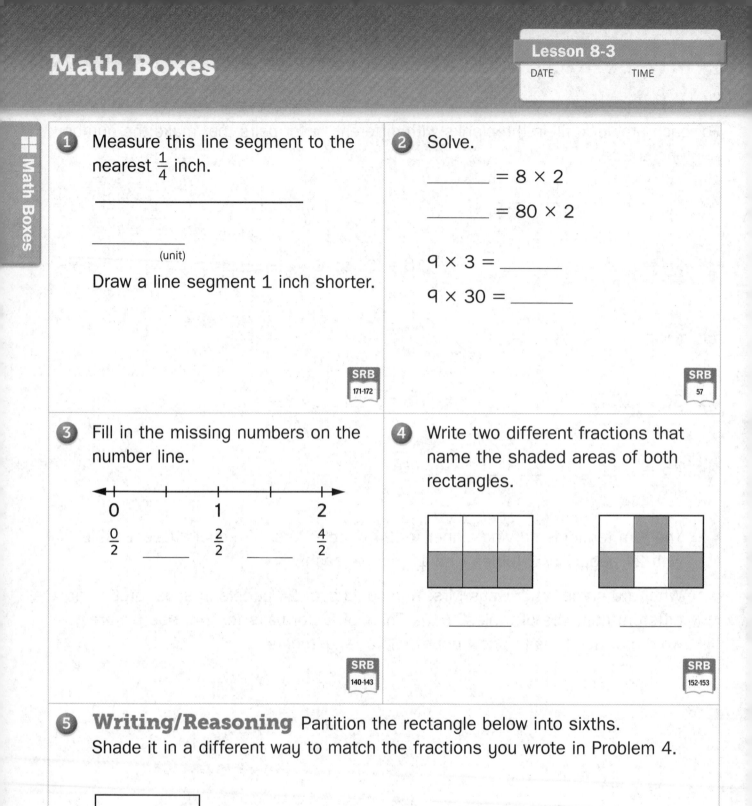

1 Measure this line segment to the nearest $\frac{1}{4}$ inch.

_____

_____
(unit)

Draw a line segment 1 inch shorter.

SRB
171-172

2 Solve.

_____ $= 8 \times 2$

_____ $= 80 \times 2$

$9 \times 3 =$ _____

$9 \times 30 =$ _____

SRB
57

3 Fill in the missing numbers on the number line.

0   1   2

$\frac{0}{2}$ _____ $\frac{2}{2}$ _____ $\frac{4}{2}$

SRB
140-143

4 Write two different fractions that name the shaded areas of both rectangles.

_____   _____

SRB
152-153

5 **Writing/Reasoning** Partition the rectangle below into sixths. Shade it in a different way to match the fractions you wrote in Problem 4.

SRB
152-153

Setting Up Chairs

Ms. Soto is setting up chairs for Math Night. Her room cannot fit more than 35 chairs. She places the same number of chairs in each row. As she sets up the chairs, she makes up a problem for her class with these clues:

Clue A: When there are 2 chairs in each row, there is 1 leftover chair.

Clue B: When there are 3 chairs in each row, there is 1 leftover chair.

Clue C: When there are 4 chairs in each row, there is still 1 leftover chair.

Clue D: When there are 5 chairs in each row, there are no leftover chairs.

Use the clues to figure out how many chairs Ms. Soto set up.

Joi, one of Ms. Soto's students, makes a conjecture that Ms. Soto set up 13 chairs.

Work with your partner and use the clues to make a mathematical argument for or against Joi's conjecture. You may draw pictures or use counters to show your thinking. Explain your reasoning.

Math Boxes

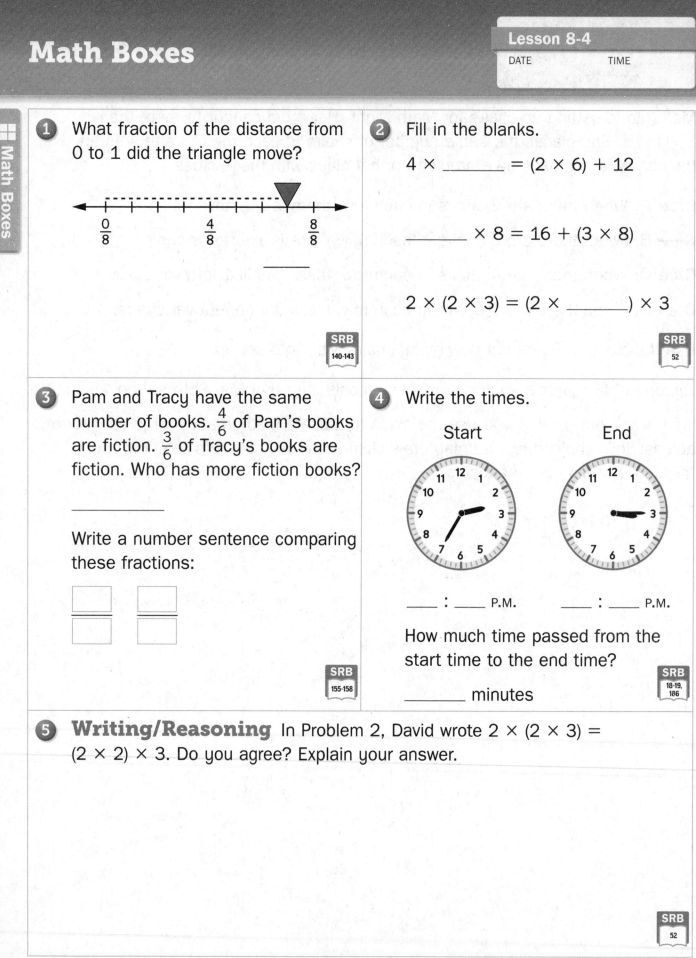

1 What fraction of the distance from 0 to 1 did the triangle move?

$\frac{0}{8}$ $\frac{4}{8}$ $\frac{8}{8}$

_____

SRB
140-143

2 Fill in the blanks.

$4 \times$ _____ $= (2 \times 6) + 12$

_____ $\times 8 = 16 + (3 \times 8)$

$2 \times (2 \times 3) = (2 \times$ _____ $) \times 3$

SRB
52

3 Pam and Tracy have the same number of books. $\frac{4}{6}$ of Pam's books are fiction. $\frac{3}{6}$ of Tracy's books are fiction. Who has more fiction books?

_____

Write a number sentence comparing these fractions:

SRB
155-158

4 Write the times.

Start   End

____ : ____ P.M.   ____ : ____ P.M.

How much time passed from the start time to the end time?

_____ minutes

SRB
18-19,
186

5 **Writing/Reasoning** In Problem 2, David wrote $2 \times (2 \times 3) = (2 \times 2) \times 3$. Do you agree? Explain your answer.

SRB
52

Factor Bingo Game Mat

Write any of the numbers 2 through 90 on the grid above.

You may use a number only once.

To help you keep track of the numbers you use, circle them in the list.

		2	3	4	5	6	7	8	9	10
11	12	13	14	15	16	17	18	19	20
21	22	23	24	25	26	27	28	29	30
31	32	33	34	35	36	37	38	39	40
41	42	43	44	45	46	47	48	49	50
51	52	53	54	55	56	57	58	59	60
61	62	63	64	65	66	67	68	69	70
71	72	73	74	75	76	77	78	79	80
81	82	83	84	85	86	87	88	89	90

Fraction Review

Complete the problems. You may use your fraction tools.

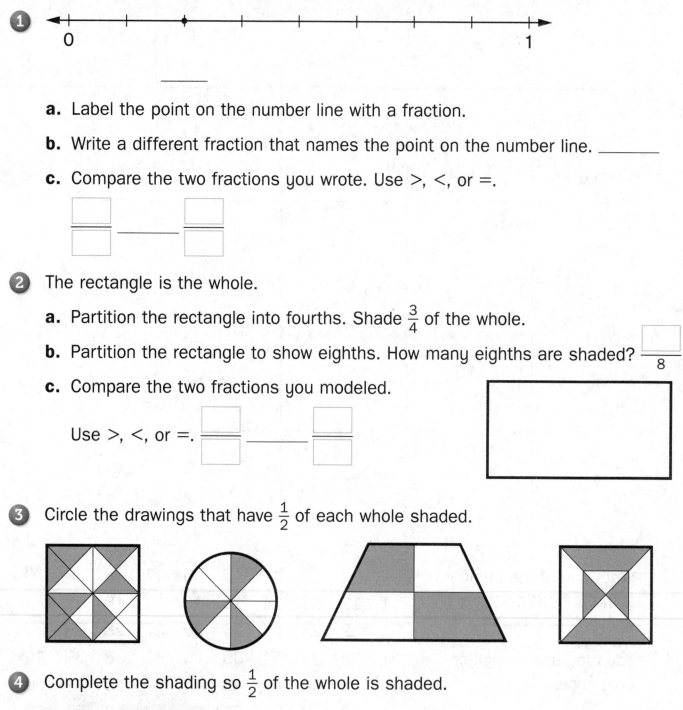

1.

    0   1

   a. Label the point on the number line with a fraction.

   b. Write a different fraction that names the point on the number line. _____

   c. Compare the two fractions you wrote. Use >, <, or =.

2. The rectangle is the whole.

   a. Partition the rectangle into fourths. Shade $\frac{3}{4}$ of the whole.

   b. Partition the rectangle to show eighths. How many eighths are shaded? $\frac{}{8}$

   c. Compare the two fractions you modeled.

      Use >, <, or =.

3. Circle the drawings that have $\frac{1}{2}$ of each whole shaded.

4. Complete the shading so $\frac{1}{2}$ of the whole is shaded.

Math Boxes

Math Boxes

1
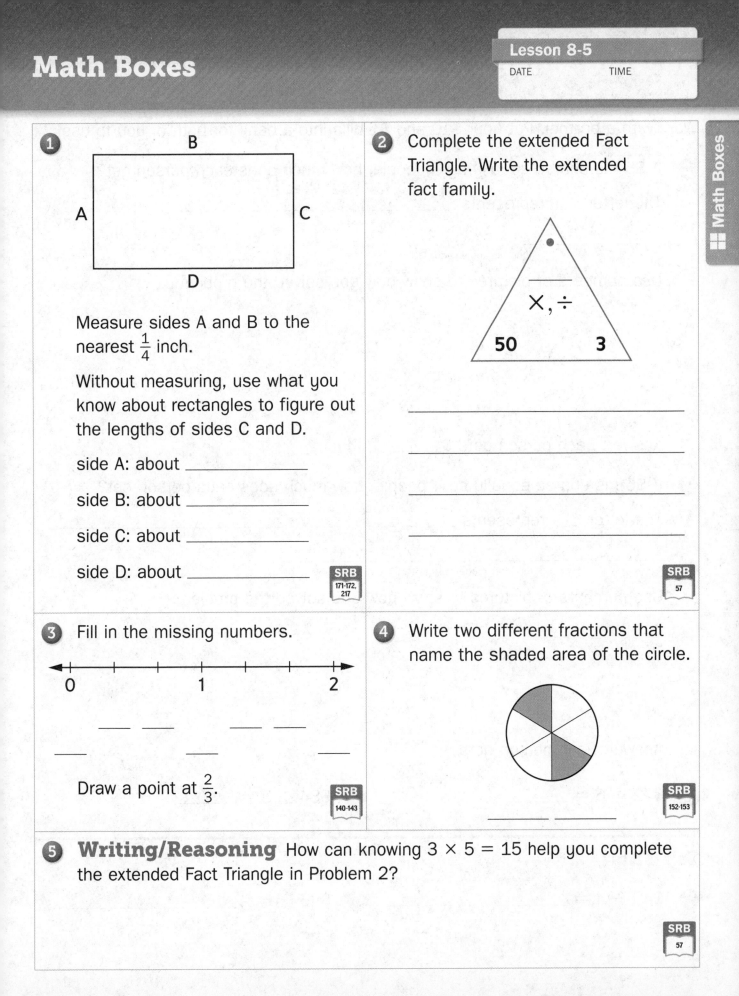

B

A   C

D

Measure sides A and B to the nearest $\frac{1}{4}$ inch.

Without measuring, use what you know about rectangles to figure out the lengths of sides C and D.

side A: about _____

side B: about _____

side C: about _____

side D: about _____

SRB
171-172, 217

2 Complete the extended Fact Triangle. Write the extended fact family.

×, ÷

50   3

_____

_____

_____

_____

SRB
57

3 Fill in the missing numbers.

0   1   2

___ ___   ___ ___

___   ___   ___

Draw a point at $\frac{2}{3}$.

SRB
140-143

4 Write two different fractions that name the shaded area of the circle.

_____ _____

SRB
152-153

5 **Writing/Reasoning** How can knowing $3 \times 5 = 15$ help you complete the extended Fact Triangle in Problem 2?

SRB
57

Sharing Money

Work with a partner. Put your $10 and $1 bills into a bank for both of you to use.

 1 If $54 is shared equally by 3 people, how much does each person get?

The letter ____ represents _____.

_____
(number model with letter)

Use numbers or pictures to show how you solved the problem.

Answer: Each person gets $_____.

2 If $68 is shared equally by 4 people, how much does each person get?

The letter ____ represents _____.

_____
(number model with letter)

Use numbers or pictures to show how you solved the problem.

Answer: Each person gets $_____.

3 $72 ÷ 6 = _____

4 $84 ÷ 3 = _____

Try This

5 $49 ÷ 4 = _____

6 $93 ÷ 6 = _____

Math Boxes

1 Shade 3 of the circles.

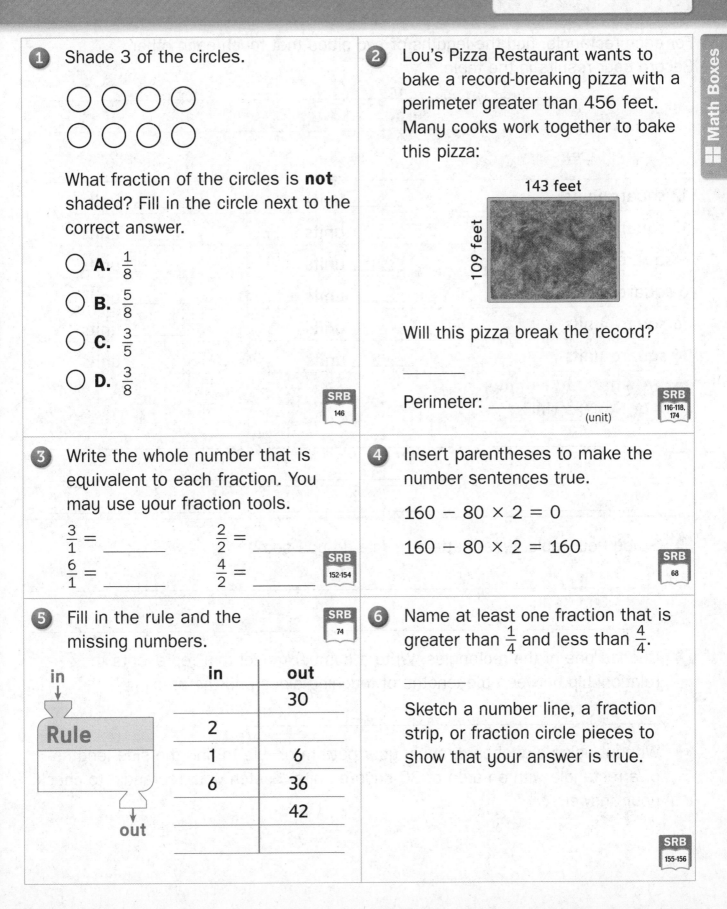

What fraction of the circles is **not** shaded? Fill in the circle next to the correct answer.

○ **A.** $\frac{1}{8}$

○ **B.** $\frac{5}{8}$

○ **C.** $\frac{3}{5}$

○ **D.** $\frac{3}{8}$

SRB
146

2 Lou's Pizza Restaurant wants to bake a record-breaking pizza with a perimeter greater than 456 feet. Many cooks work together to bake this pizza:

143 feet

109 feet

Will this pizza break the record?

_____

Perimeter: _____
                    (unit)

SRB
116-118,
174

3 Write the whole number that is equivalent to each fraction. You may use your fraction tools.

$\frac{3}{1} =$ _____ $\frac{2}{2} =$ _____

$\frac{6}{1} =$ _____ $\frac{4}{2} =$ _____

SRB
152-154

4 Insert parentheses to make the number sentences true.

$160 - 80 \times 2 = 0$

$160 - 80 \times 2 = 160$

SRB
68

5 Fill in the rule and the missing numbers.

SRB
74

in
↓

Rule

↓
out

in	out
	30
2	
1	6
6	36
	42

6 Name at least one fraction that is greater than $\frac{1}{4}$ and less than $\frac{4}{4}$.

_____

Sketch a number line, a fraction strip, or fraction circle pieces to show that your answer is true.

SRB
155-156

Exploring Geoboard Areas

For each rectangle, find the lengths of two sides that touch each other.
Record your results in the table.

Geoboard Areas		
Area	**Side 1**	**Side 2**
12 square units	_____ units	_____ units
12 square units	_____ units	_____ units
6 square units	_____ units	_____ units
6 square units	_____ units	_____ units
16 square units	_____ units	_____ units
16 square units	_____ units	_____ units
Area with Odd Number of Square Units	**Side 1**	**Side 2**
_____	_____	_____
_____	_____	_____
_____	_____	_____

1. Study your table. What pattern or rule do you see?

   _____

   _____

2. Choose one of the rectangles. Write a number model that represents the
   relationship between the lengths of touching sides and the area.

   _____

3. Without using a geoboard, apply your pattern or rule to find the side lengths
   of a rectangle with an area of 30 square units. Sketch your rectangle to check
   your answer.

Exploring Equivalent Fractions

Follow the "first time you work on this card" directions on Activity Card 92.

The red circle is the whole.

1

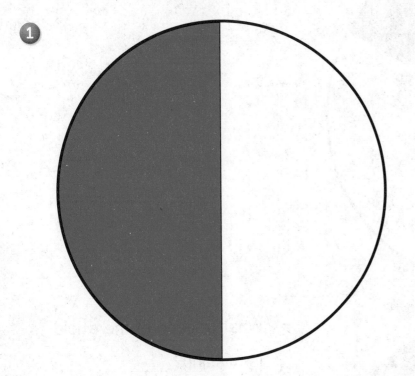

What fraction of the whole

is missing? _____

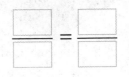

2

What fraction of the whole

is missing? _____

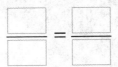

3

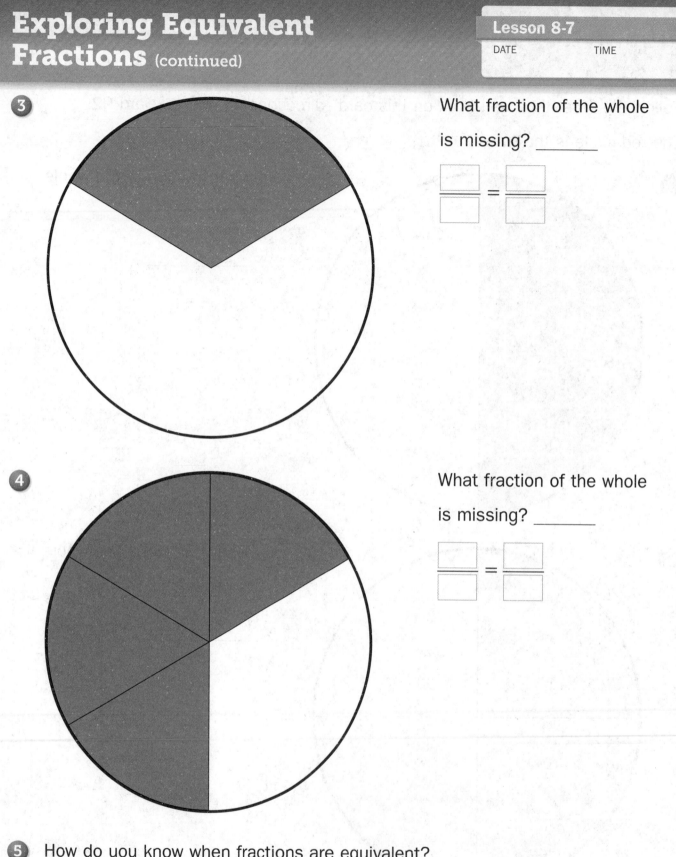

What fraction of the whole is missing? _____

$$\frac{\boxed{}}{\boxed{}} = \frac{\boxed{}}{\boxed{}}$$

4

What fraction of the whole is missing? _____

$$\frac{\boxed{}}{\boxed{}} = \frac{\boxed{}}{\boxed{}}$$

5 How do you know when fractions are equivalent?

_____

_____

Math Boxes
Preview for Unit 9

1 On the first day of spring, the length of daytime and nighttime are about the same. There are 24 hours in a day. What is the length of daytime?

about _____
(unit)

If the sun rises at 6:51 A.M. on that day, at about what time would you expect it to set?

about _____ P.M.

SRB
18-19,
187-188

2 Put these facts in order from the least to the greatest product.

7 × 6   8 × 6   7 × 5   6 × 5   4 × 2

least   _____

_____

_____

_____

greatest _____

SRB
54

3 Solve.

5 × 60 = _____

70 × _____ = 280

_____ × 5 = 450

6 × _____ = 360

SRB
57

4 Fill in the circles next to all the ways to break apart 14 × 6.

Ⓐ 10 × 6 + 4 × 6

Ⓑ 7 × 6 + 7 × 6

Ⓒ 6 × 6 + 6 × 6

Ⓓ 2 × 6 + 12 × 6

SRB
49, 51

5

Books Sold at Book Fair

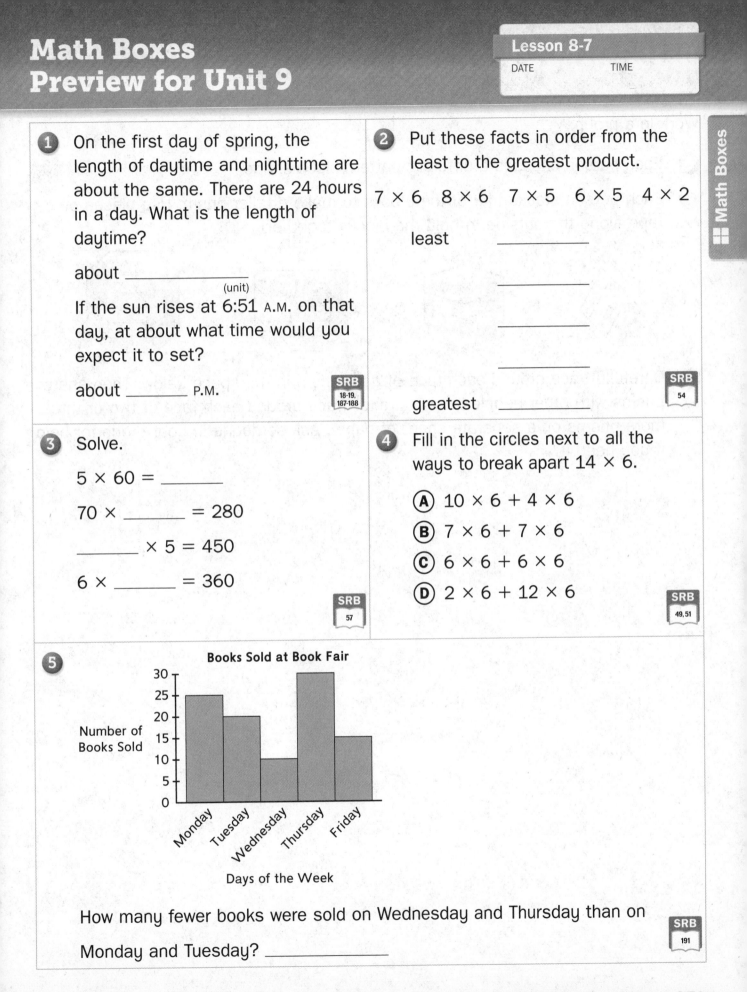

How many fewer books were sold on Wednesday and Thursday than on Monday and Tuesday? _____

SRB
191

Math Boxes

Pattern-Block Prisms

Work in a group.

1 Each person chooses a different pattern-block shape.

2 Each person stacks four of the blocks to make a taller prism. Use pieces of tape along the outside to hold the blocks together.

  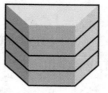 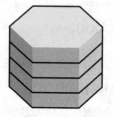

3 Carefully trace around each face of your prism in the space below. Then share prisms with other people in your group. Trace around each face of two or three more prisms on a separate sheet of paper. Ask someone in your group for help if you need it.

Math Boxes

1 Draw a square with $1\frac{1}{4}$-inch sides.

SRB
171-172,
217

2 Complete the extended Fact Triangle. Write the extended fact family.

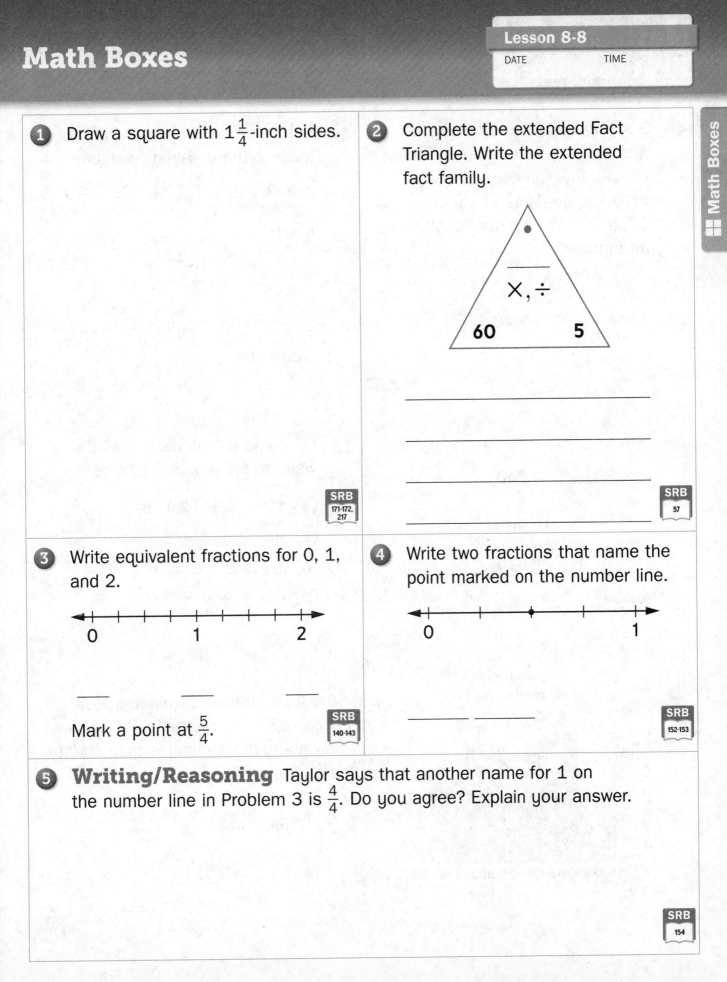

×, ÷

60   5

_____

_____

_____

_____

SRB
57

3 Write equivalent fractions for 0, 1, and 2.

0   1   2

____   ____   ____

Mark a point at $\frac{5}{4}$.

SRB
140-143

4 Write two fractions that name the point marked on the number line.

0   1

____   ____

SRB
152-153

5 **Writing/Reasoning** Taylor says that another name for 1 on the number line in Problem 3 is $\frac{4}{4}$. Do you agree? Explain your answer.

SRB
154

Math Boxes
Preview for Unit 9

1 Summer days are very long in Anchorage, Alaska. In the middle of July, the sun rises at about 3:20 A.M. and sets at about 10:20 P.M. What is the length of daytime?

about _____ hours

SRB
18-19,
187-188

2 Put these facts in order from the least to the greatest product.

4×9 8×5 9×5 9×3

least   _____

_____

_____

greatest   _____

SRB
54

3 Solve.

$8 \times$ _____ $= 240$

$4 \times 90 =$ _____

_____ $\times 7 = 420$

$70 \times$ _____ $= 560$

SRB
57

4 Fill in the circles next to all the ways to break apart 18×3.

(A) $12 \times 3 + 12 \times 3$

(B) $9 \times 3 + 2 \times 3$

(C) $10 \times 3 + 8 \times 3$

(D) $9 \times 3 + 9 \times 3$

SRB
49, 51

5

Minutes Spent at the Park

Find the total number of minutes Jack and Dylan spent at the park. Then figure out how many more minutes they spent at the park than Eve.

_____
                    (unit)

SRB
191

Reviewing Area and Perimeter

In the spring, the Garden Club will plant a garden. Each child will have 1 square meter of the garden to plant. There are 16 children in the club, so the area of the garden will be 16 square meters. Four children drew shapes for the garden below.

1 Find the area of each shape. Circle the shapes that have an area of 16 square meters. Cross out the shapes that do not have an area of 16 square meters.

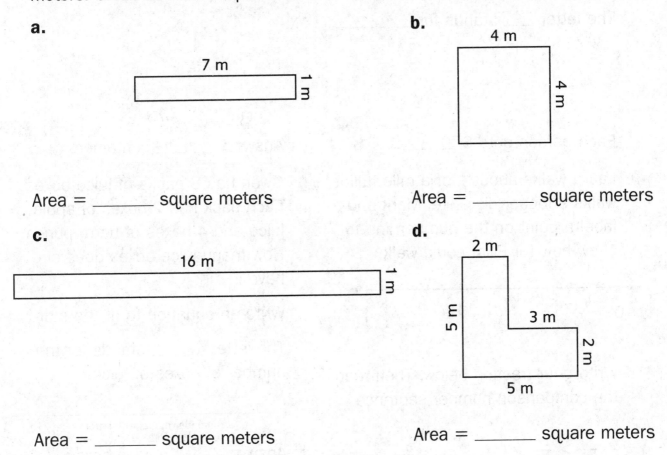

a.

7 m

1 m

Area = _____ square meters

b.

4 m

4 m

Area = _____ square meters

c.

16 m

1 m

Area = _____ square meters

d.

2 m

5 m

3 m

2 m

5 m

Area = _____ square meters

2 The club wants to build a fence around the garden, but they do not want to spend a lot of money. They need to find a shape that has an area of 16 square meters and the shortest perimeter.

Which of the above shapes has an area of 16 square meters and the

shortest perimeter? _____

How did you find the perimeter of this shape? _____

_____

_____

Math Boxes

1 Share $42 equally among 3 people. You may use $10 bills and $1 bills.

_____
(number model with letter)

The letter ____ stands for _____.

Each person gets $_____.

SRB
40

2 Each juice container holds 3-liters. How many containers are needed for 27 liters of juice?

Answer: _____ containers

SRB
182

3 Laura walks about $\frac{1}{6}$ of a mile. Elliot walks a longer distance. Mark and label a point on the number line to show how far Elliot could walk.

0      $\frac{1}{6}$ 1

Write your fraction below. Then read the comparison number sentence.

$\dfrac{\boxed{}}{\boxed{}} > \frac{1}{6}$

SRB
144,
157-158

4 Gwen has 8 packs of juice boxes. Each pack has 2 boxes of apple juice and 4 boxes of berry punch. How many juice boxes does she have in all?

Write an equation to fit the story.

The letter _____ stands for the number of boxes of juice.

_____
(number model with letter)

Answer: _____ juice boxes

SRB
61-64

5 **Writing/Reasoning** Explain how you plotted a distance that Elliot could walk in Problem 3.

SRB
144

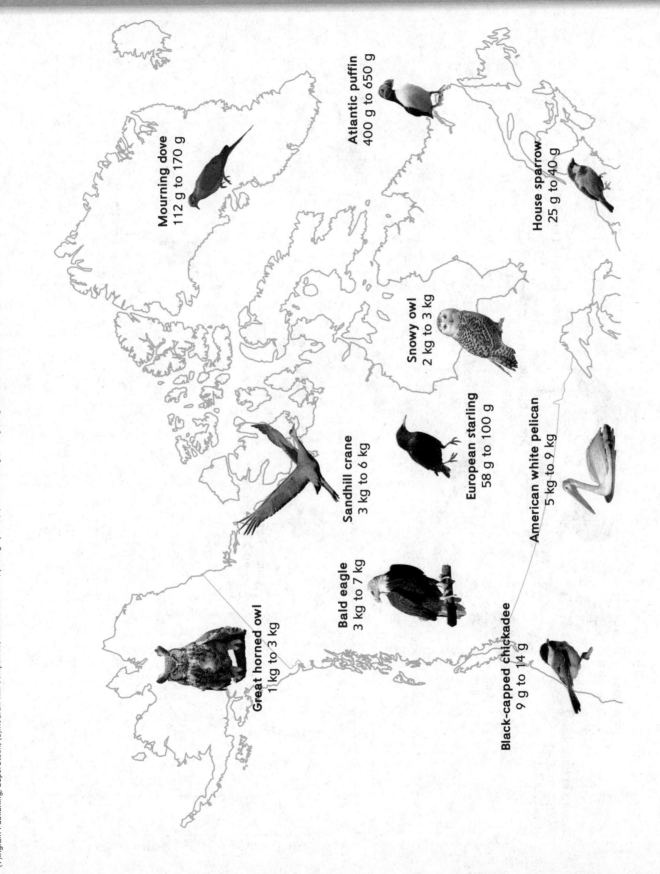

Atlantic puffin
400 g to 650 g

Mourning dove
112 g to 170 g

House sparrow
25 g to 40 g

Snowy owl
2 kg to 3 kg

Sandhill crane
3 kg to 6 kg

European starling
58 g to 100 g

American white pelican
5 kg to 9 kg

Great horned owl
1 kg to 3 kg

Bald eagle
3 kg to 7 kg

Black-capped chickadee
9 g to 14 g

(l to r, t to b)McGraw-Hill Education/Brian Kanof, (2)©imagebroker/Alamy, (3)Andrew Howe/Getty Images, (4)Enjoylife2/iStock/360/Getty Images, (5)Elemental Imaging/Getty Images, (6)Charlie Bishop/Getty Images, (7)Ingram Publishing/SuperStock, (8)McGraw-Hill Companies, Inc. Mark Dierker, photographer, (9)Robbie George/Getty Images, (10)Jonathan Larsen/Diadem Images/Alamy

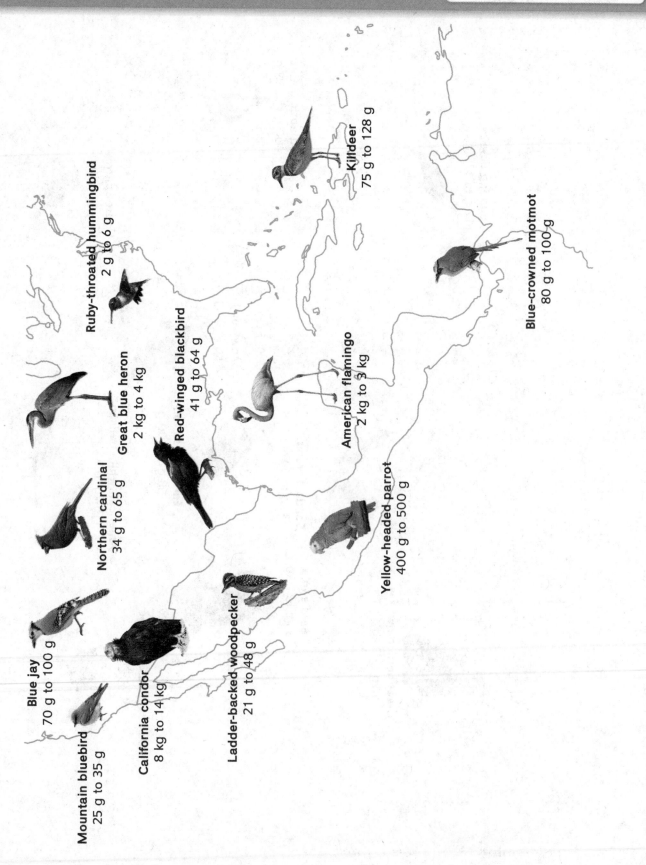

Killdeer
75 g to 128 g

Ruby-throated hummingbird
2 g to 6 g

Blue-crowned motmot
80 g to 100 g

Great blue heron
2 kg to 4 kg

Red-winged blackbird
41 g to 64 g

American flamingo
2 kg to 3 kg

Northern cardinal
34 g to 65 g

Blue jay
70 g to 100 g

Yellow-headed parrot
400 g to 500 g

Mountain bluebird
25 g to 35 g

California condor
8 kg to 14 kg

Ladder-backed woodpecker
21 g to 48 g

(l to r, t to b)Charles Bruttag/Getty Images, (2)©Daniel Dempster graphy/Alamy, (3)Kelley Miller/Getty Images, (4)©Brand X Pictures/PunchStock, (5)©Corbis/PunchStock, (6)©James Urbach/SuperStock, (7)John Anderson/Getty Images, (8)Powerofforever/iStock/Getty Images Plus/Getty Images, (9)©James Urbach/SuperStock, (10)kojihirano/Getty Images, (11)MikeLane45/Getty Images, (12)Design Pics/Philippe Widling

North American Bird Number Stories

Solve each number story. You may draw a picture or a multiplication/division diagram.

1. What is the total mass of 70 bald eagles that each have a mass of about 4 kilograms?

   Number model: _____

   Answer: about _____ kg

2. What bird could have the same mass as three 30-gram ladder-backed woodpeckers?

   Number model: _____

   Answer: _____

3. There are 6 birds that each have about the same mass taking baths in a birdbath. Their combined mass is about 180 grams. Name at least two types of bird they could be.

   Number model: _____

   Answer: _____

Math Boxes

① Fill in the blanks.

12 = _____ × _____

12 = _____ × _____

12 = _____ × _____

SRB
65-66

② $1 = \dfrac{\boxed{}}{4}$

Partition this number line into fourths. Label $\frac{1}{4}$, $\frac{2}{4}$, $\frac{3}{4}$, and $\frac{4}{4}$.

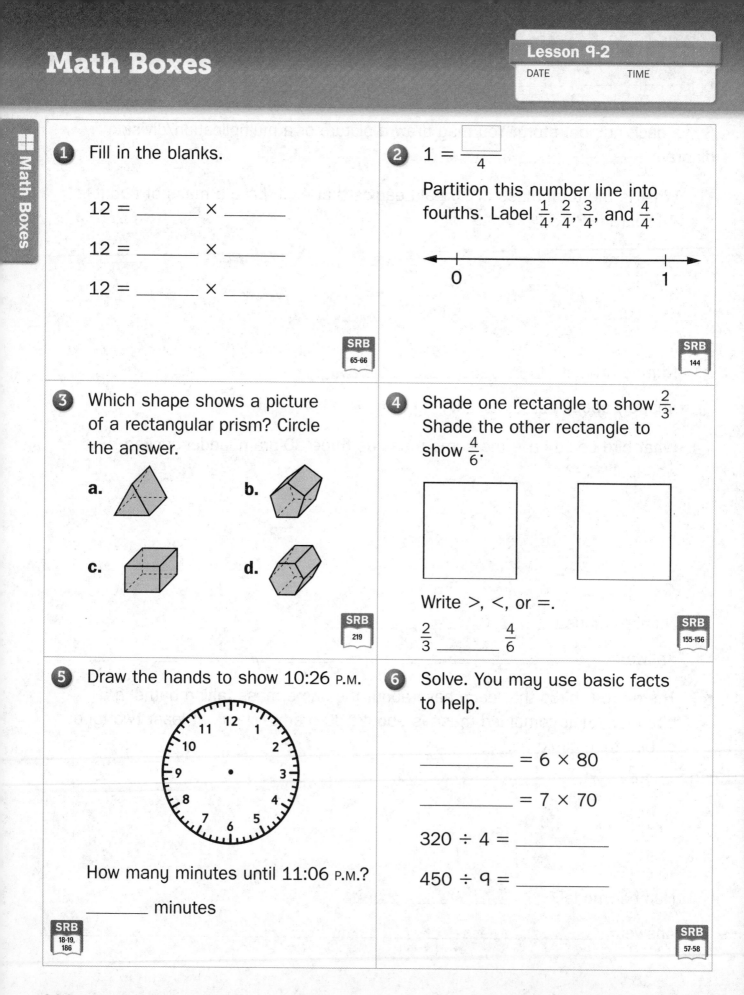

0   1

SRB
144

③ Which shape shows a picture of a rectangular prism? Circle the answer.

a.

b.

c.

d.

SRB
219

④ Shade one rectangle to show $\frac{2}{3}$. Shade the other rectangle to show $\frac{4}{6}$.

Write >, <, or =.

$\dfrac{2}{3}$ _____ $\dfrac{4}{6}$

SRB
155-156

⑤ Draw the hands to show 10:26 P.M.

How many minutes until 11:06 P.M.?

_____ minutes

SRB
18-19,
186

⑥ Solve. You may use basic facts to help.

_____ = 6 × 80

_____ = 7 × 70

320 ÷ 4 = _____

450 ÷ 9 = _____

SRB
57-58

Using Mental Math to Multiply

Solve each number story in your head. Use number models and words to show your thinking.

1 The mass of one American white pelican is about 8 kilograms. What is the mass of sixteen 8-kilogram pelicans?

My thinking: _____

_____

Answer: about _____ kilograms

2 One blue jay has a mass of about 75 grams. What bird could have about the same mass as six 75-gram blue jays?

My thinking: _____

_____

Answer: _____

3 One house sparrow has a mass of about 35 grams. How many blue-crowned motmots together would have about the same mass as five 35-gram sparrows?

My thinking: _____

_____

Answer: _____ blue-crowned motmots

Math Boxes

1 Share $57 equally among 3 people. You may use $10 bills and $1 bills.

_____
(number model with letter)

The letter _____ stands for _____.

Each person gets $_____.

SRB
40

2 Martin poured 24 liters of water into 8 same-sized containers. How many liters of water did each container hold?

Answer: _____ liters

SRB
182

3 Sierra walks $\frac{6}{8}$ of a mile to school. Abby walks a shorter distance. Complete the number sentence to show what fraction of a mile Abby could walk.

$$\frac{\boxed{}}{\boxed{}} < \frac{6}{8}$$

SRB
157-158

4 Martina has 8 cans with 3 tennis balls each. Andre borrows 4 cans for practice. How many tennis balls does Martina have left?

The letter _____ stands for the number of tennis balls left.

_____
(number model with letter)

Martina has _____ balls left.

SRB
61-64

5 **Writing/Reasoning** Look at Problem 1. If 4 people share $57, would they get more or less money than 3 people who share $57? How do you know?

Planning a Field Trip

Use Activity Card 98 and fill out the schedule for Ms. Bernstein's third-grade class to follow at the City Zoo. You may use your toolkit clock or draw an open number line at the bottom of this page.

Schedule for a Field Trip to the City Zoo		
Start Time	**End Time**	**Activity**

Building Bridges Record Sheet

Bridge	Mass (in grams)
1	_____ grams
2	_____ grams
3	_____ grams

1 What is the difference (in grams) between what the strongest bridge can hold and what the weakest bridge can hold?

_____

_____

2 How is Bridge 2 different from Bridge 1?

_____

_____

3 How does the size of the folds affect how much mass the bridge holds?

_____

_____

4 Why do you think the strongest bridge can hold more mass?

_____

_____

Math Boxes

1 Circle all of the products below that have 4 as a factor.

   A. 24

   B. 14

   C. 12

   D. 28

SRB
65-66

2 Label $\frac{1}{4}$ and $\frac{1}{8}$ on the number line.

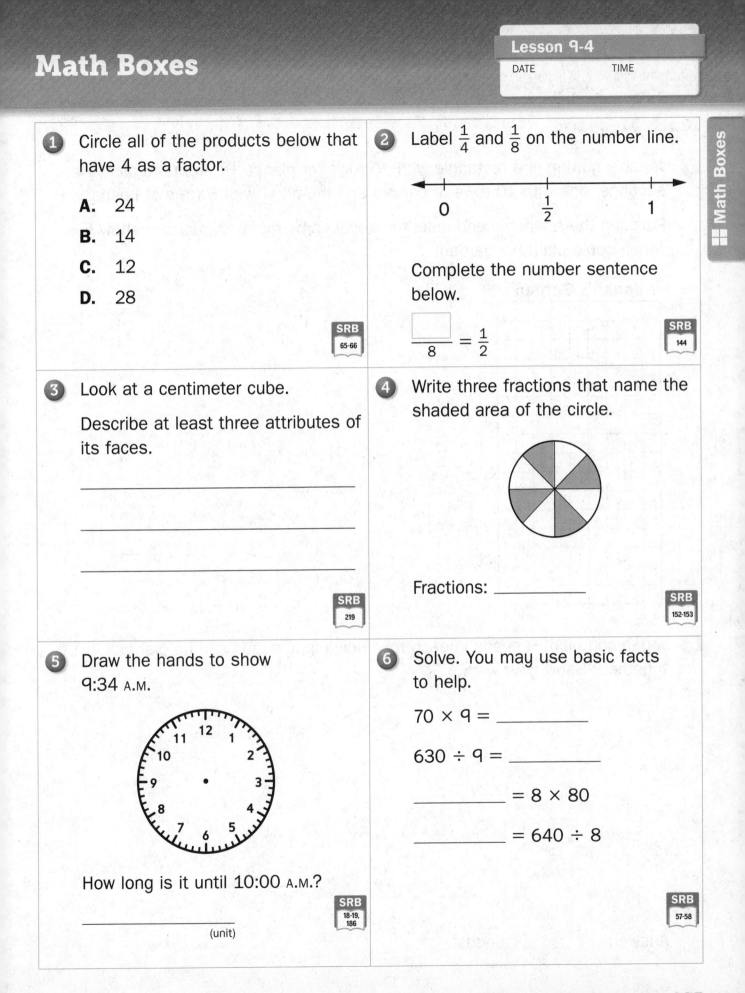

0         $\frac{1}{2}$ 1

Complete the number sentence below.

$$\frac{\boxed{}}{8} = \frac{1}{2}$$

SRB
144

3 Look at a centimeter cube.

Describe at least three attributes of its faces.

_____

_____

_____

SRB
219

4 Write three fractions that name the shaded area of the circle.

Fractions: _____

SRB
152-153

5 Draw the hands to show 9:34 A.M.

How long is it until 10:00 A.M.?

_____
(unit)

SRB
18-19,
186

6 Solve. You may use basic facts to help.

70 × 9 = _____

630 ÷ 9 = _____

_____ = 8 × 80

_____ = 640 ÷ 8

SRB
57-58

Jonah's Garden

Math Message

1   Jonah's garden is a rectangle with 16 rows for plants. He wants to plant two sections: one with 10 rows of carrots and the other with 6 rows of beans.

Partition the rectangle and label the sections *carrots* and *beans* to show how Jonah could plant his garden.

Jonah's Garden

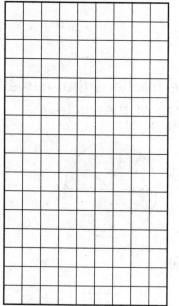

2   Jonah can plant 9 seeds in each row. How many seeds can he plant all together? Show your work.

Answer: _____ seeds

Decomposing Factors to Multiply

Use the break-apart strategy to solve the multiplication problems. Think of easier multiplication problems that you can use to break apart the larger factor. Draw rectangles and write number sentences to show your thinking.

1 $7 \times 24 =$ _____

I can break _____ into _____.

2 $4 \times 36 =$ _____

I can break _____ into _____.

3 $8 \times 47 =$ _____

I can break _____ into _____.

Math Boxes

1 The total mass of 10 same-size ruby-throated hummingbirds is about 60 grams. What is the mass of one hummingbird? You may fill in the diagram to help.

number of hummingbirds	grams per hummingbird	grams in all

About _____ grams

SRB 63-64

2 Shade the beakers to show $\frac{1}{2}$ liter of water.

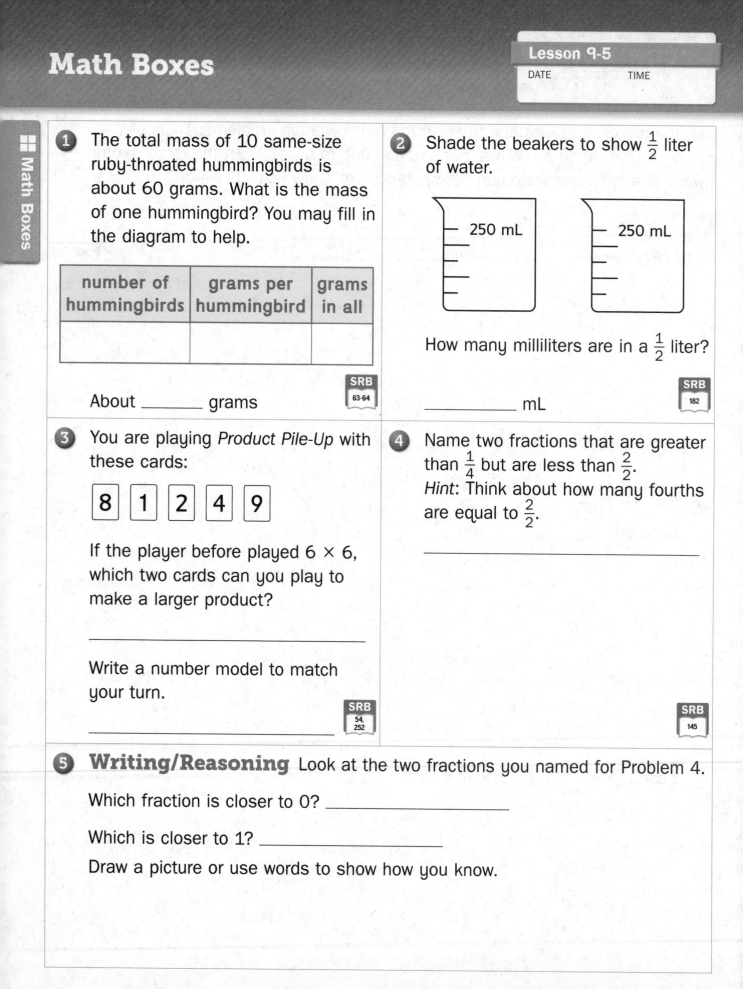

| 250 mL | 250 mL |

How many milliliters are in a $\frac{1}{2}$ liter?

_____ mL

SRB 182

3 You are playing *Product Pile-Up* with these cards:

| 8 | 1 | 2 | 4 | 9 |

If the player before played 6 × 6, which two cards can you play to make a larger product?

_____

Write a number model to match your turn.

_____

SRB 54, 252

4 Name two fractions that are greater than $\frac{1}{4}$ but are less than $\frac{2}{2}$.

Hint: Think about how many fourths are equal to $\frac{2}{2}$.

_____

SRB 145

5 **Writing/Reasoning** Look at the two fractions you named for Problem 4.

Which fraction is closer to 0? _____

Which is closer to 1? _____

Draw a picture or use words to show how you know.

Working with a Broken Calculator

Math Message

Pretend that your calculator is broken. The addition key ($+$) and the number 8 key (8) are not working. Think about how you could use your broken calculator to make the number 18.

In the name-collection box, write at least five different names for 18 that do not use the addition key ($+$) or the number 8 key (8). Use two operations ($-$, $\times$, $\div$) in at least one of the names.

18

Math Boxes

1 Measure to the nearest $\frac{1}{4}$ inch and label the sides of this quadrilateral.

Another name for this quadrilateral is a _____.

SRB
171-172,
216-217

2 There are 8 tulips. Tanya puts $\frac{2}{8}$ of the tulips into each bouquet of flowers. How many bouquets can she make?

_____ bouquets

SRB
57-58

3 Circle the facts that have a product greater than 20.

A. 5 × 6

B. 4 × 4

C. 8 × 2

D. 9 × 3

SRB
54

4 What is the mass of fifteen 8 kg California condors? Solve. Record number models to show your thinking.

_____
(number models)

Answer: _____

(unit)

SRB
59-60

5 Solve.

4 × 60 = _____

_____ = 80 × 9

250 ÷ 5 = _____

_____ = 490 ÷ 7

SRB
146-147

6 Use the order of operations to solve.

240 + (6 × 60) = _____

_____ = (645 − 605) × 4

SRB
69-70

Math Boxes

Calculating the Length of Day

Use open number lines, a toolkit clock, or other representations to help you calculate the elapsed times. Use the sunrise and sunset data on *Student Reference Book,* page 281.

 1 Find the length of day for Esperanza Base, Antarctica, on June 21, 2016. Show your work.

Sunrise time: _____ Sunset time: _____

Elapsed time: _____

2 Find the length of day for another location on the map. Show your work.

Location: _____

Sunrise time: _____ Sunset time: _____

Elapsed time: _____

Math Boxes

1 The total mass of a flock of chickadees is 90 grams. Each chickadee has a mass of about 10 grams. How many chickadees are in the flock? You may fill in the diagram to help.

number of chickadees	grams per chickadee	grams in flock

_____
(unit)

SRB
63-64

2

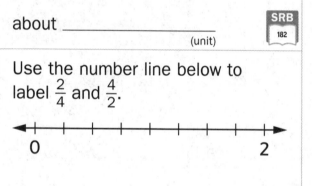

If Deena pours out 250 mL, how much water will be left?

about _____
(unit)

SRB
182

3 You have these cards while playing *Product Pile-Up.*

| 0 | 8 | 5 | 10 | 2 | 4 | 9 | 4 |

Which cards would you play if you get to go first?

_____

Write a number sentence to match your turn.

_____

SRB
54,
252

4 Use the number line below to label $\frac{2}{4}$ and $\frac{4}{2}$.

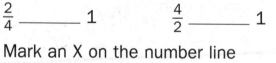
0   2

Write <, >, or = to make each number sentence true.

$\frac{2}{4}$ _____ 1 $\frac{4}{2}$ _____ 1

Mark an X on the number line to show which fraction is closer to 1.

SRB
145

5 **Writing/Reasoning** Explain how you used the number line to figure out which fraction is closer to 1 in Problem 4.

SRB
140

Math Boxes

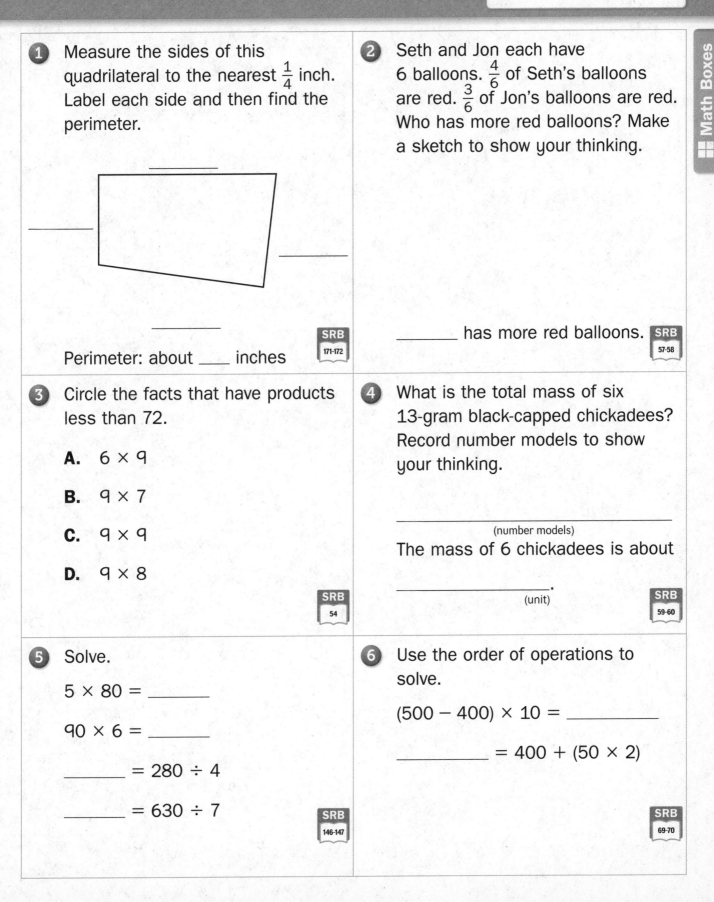

1 Measure the sides of this quadrilateral to the nearest $\frac{1}{4}$ inch. Label each side and then find the perimeter.

Perimeter: about ____ inches

SRB 171-172

2 Seth and Jon each have 6 balloons. $\frac{4}{6}$ of Seth's balloons are red. $\frac{3}{6}$ of Jon's balloons are red. Who has more red balloons? Make a sketch to show your thinking.

_____ has more red balloons.

SRB 57-58

3 Circle the facts that have products less than 72.

A. 6×9

B. 9×7

C. 9×9

D. 9×8

SRB 54

4 What is the total mass of six 13-gram black-capped chickadees? Record number models to show your thinking.

_____
(number models)

The mass of 6 chickadees is about

_____.
(unit)

SRB 59-60

5 Solve.

$5 \times 80 =$ _____

$90 \times 6 =$ _____

_____ $= 280 \div 4$

_____ $= 630 \div 7$

SRB 146-147

6 Use the order of operations to solve.

$(500 - 400) \times 10 =$ _____

_____ $= 400 + (50 \times 2)$

SRB 69-70

My Multiplication Facts Strategy Log 7

Strategy: _____

Example:

$8 \times 6 = ?$

Helper fact: $4 \times 6 = 24$

I double 4×6 because 8 is 4 doubled.

$24 + 24 = 48$

$8 \times 6 = 48$

Sketch:

```
┌──────────────┐
│              │
│  4 × 6 = 24  │
│              │ ⎫
├──────────────┤ ⎬ 8 feet
│              │ ⎭
│  4 × 6 = 24  │
│              │
└──────────────┘
      6 feet
```

My examples:

Fact: _____ × _____ = _____

Sketch:

Helper fact: _____ × _____ = _____

Fact: _____ × _____ = _____

Sketch:

Helper fact: _____ × _____ = _____

My Multiplication Facts Strategy Log 8

Strategy: _____

Example:

$8 \times 7 = ?$

Helper fact: $7 \times 7 = 49$

I start with 7 groups of 7.

I add 1 group of 7 to find 8 groups of 7.

$8 \times 7 = 56$

Sketch:

```
X X X X X X X
X X X X X X X
X X X X X X X
X X X X X X X
X X X X X X X
X X X X X X X
X X X X X X X
O O O O O O O
```

My examples:

Fact: _____ × _____ = _____

Helper fact: _____ × _____ = _____

Sketch:

Fact: _____ × _____ = _____

Helper fact: _____ × _____ = _____

Sketch:

Strategy: _____

Example:

$7 \times 6 = ?$

Helper facts: 5×6 and 2×6

I think of a 7-by-6 rectangle. I partition it into two smaller rectangles that are 5-by-6 and 2-by-6 because 5×6 and 2×6 are facts I know.

$$7 \times 6 = 5 \times 6 + 2 \times 6$$
$$= 30 + 12$$
$$= 42$$

So $7 \times 6 = 42$.

Sketch:

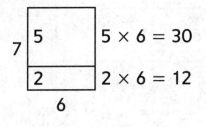

$5 \times 6 = 30$

$2 \times 6 = 12$

My examples:

Fact: _____ × _____ = _____

Helper fact: _____ × _____ = _____

Sketch:

Fact: _____ × _____ = _____

Helper fact: _____ × _____ = _____

Sketch:

My Multiplication Facts Inventory Part 1

Multiplication Fact	Know It	Don't Know It	How I Can Figure It Out
2 × 2			
10 × 5			
3 × 2			
2 × 7			
5 × 5			
9 × 2			
3 × 10			
2 × 5			
2 × 8			
5 × 4			
6 × 2			
3 × 5			
10 × 2			
2 × 4			
10 × 10			

My Multiplication Facts Inventory Part 2

Multiplication Fact	Know It	Don't Know It	How I Can Figure It Out
5 × 6			
10 × 9			
1 × 0			
9 × 5			
10 × 4			
1 × 1			
2 × 1			
8 × 10			
7 × 5			
0 × 2			
5 × 8			
10 × 6			
5 × 1			
0 × 4			
7 × 10			

My Multiplication Facts Inventory Part 3

Multiplication Fact	Know It	Don't Know It	How I Can Figure It Out
3 × 3			
1 × 4			
3 × 7			
0 × 3			
9 × 9			
6 × 0			
4 × 4			
3 × 6			
7 × 1			
6 × 6			
4 × 3			
8 × 8			
8 × 3			
3 × 9			
7 × 7			

My Multiplication Facts Inventory Part 4

Multiplication Fact	Know It	Don't Know It	How I Can Figure It Out
9 × 4			
6 × 9			
4 × 7			
8 × 6			
7 × 9			
6 × 4			
8 × 7			
6 × 4			
8 × 9			
4 × 8			

Sunrise and Sunset Data

Date	Time of Sunrise	Time of Sunset	Length of Day	
			hr	min
			hr	min
			hr	min
			hr	min
			hr	min
			hr	min
			hr	min
			hr	min
			hr	min
			hr	min
			hr	min
			hr	min
			hr	min
			hr	min
			hr	min
			hr	min
			hr	min
			hr	min
			hr	min
			hr	min
			hr	min
			hr	min
			hr	min
			hr	min
			hr	min
			hr	min

Notes

Notes

Notes

Notes

Fraction Cards 1

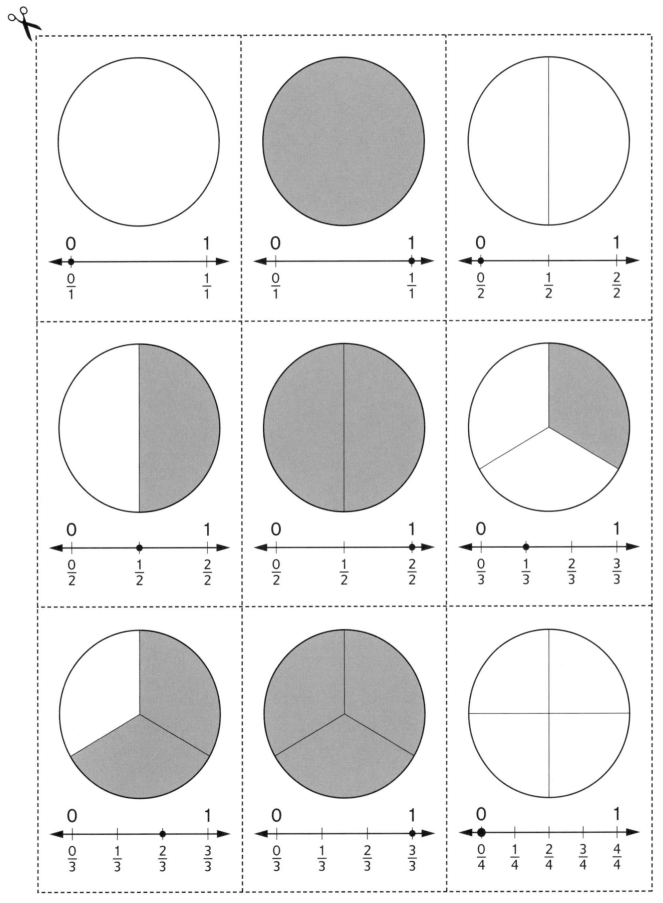

Fraction Cards 1 (continued)

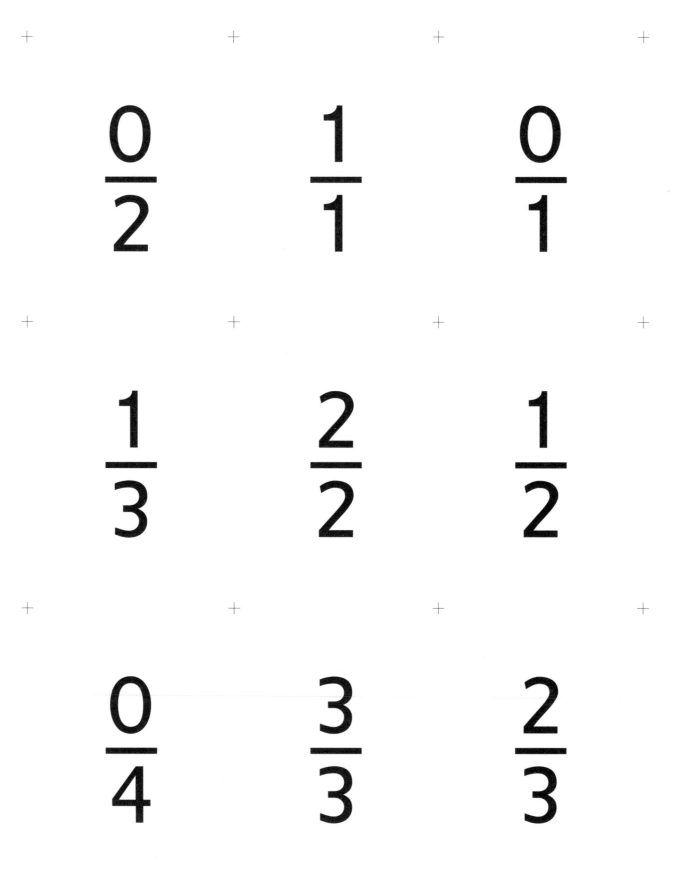

$$\frac{0}{2} \qquad \frac{1}{1} \qquad \frac{0}{1}$$

$$\frac{1}{3} \qquad \frac{2}{2} \qquad \frac{1}{2}$$

$$\frac{0}{4} \qquad \frac{3}{3} \qquad \frac{2}{3}$$

Back of Activity Sheet 16

Fraction Cards 2

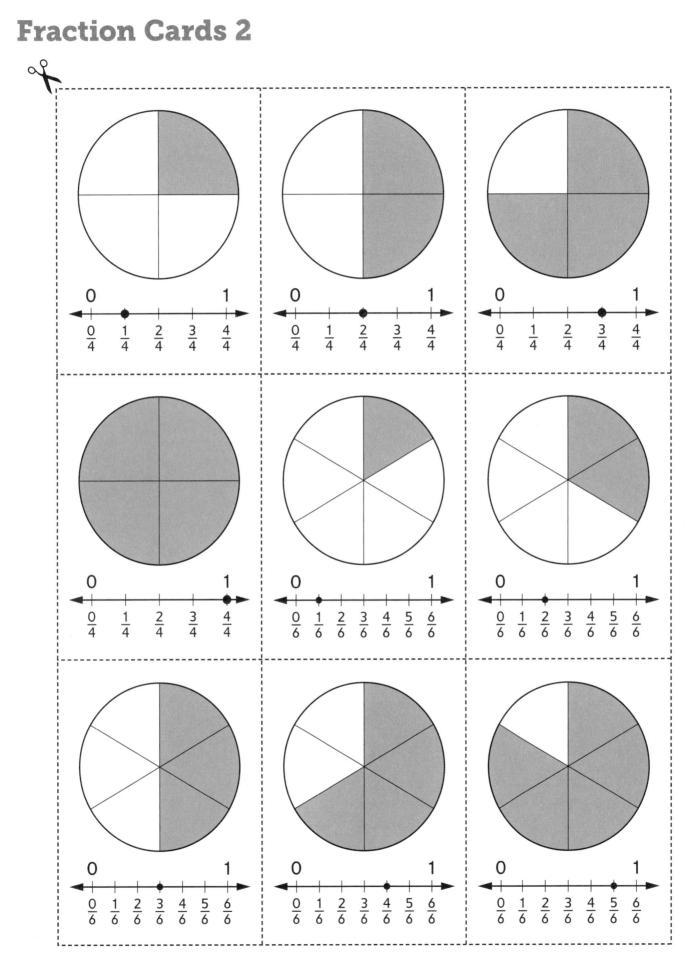

Fraction Cards 2 (continued)

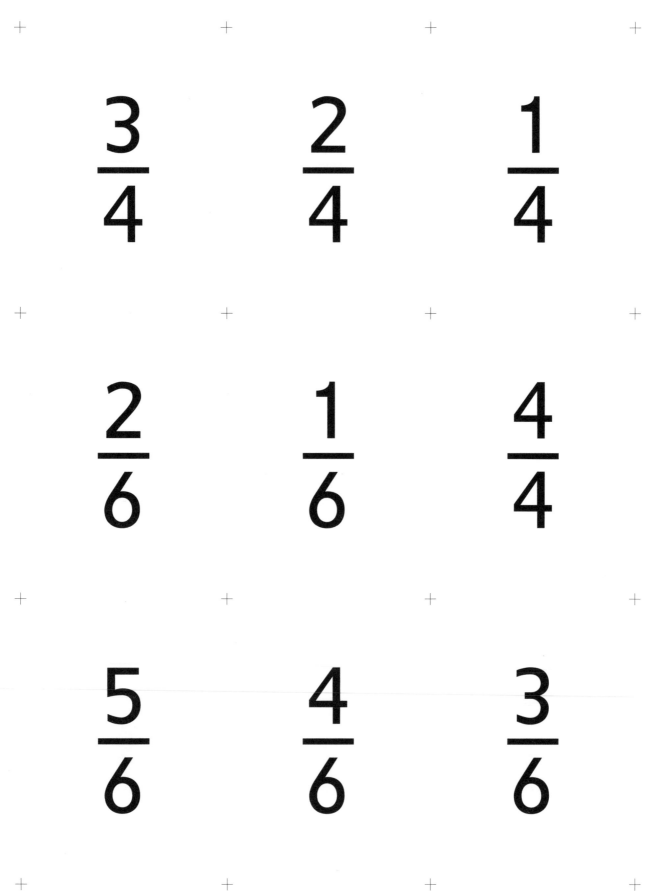

Fraction Cards 3

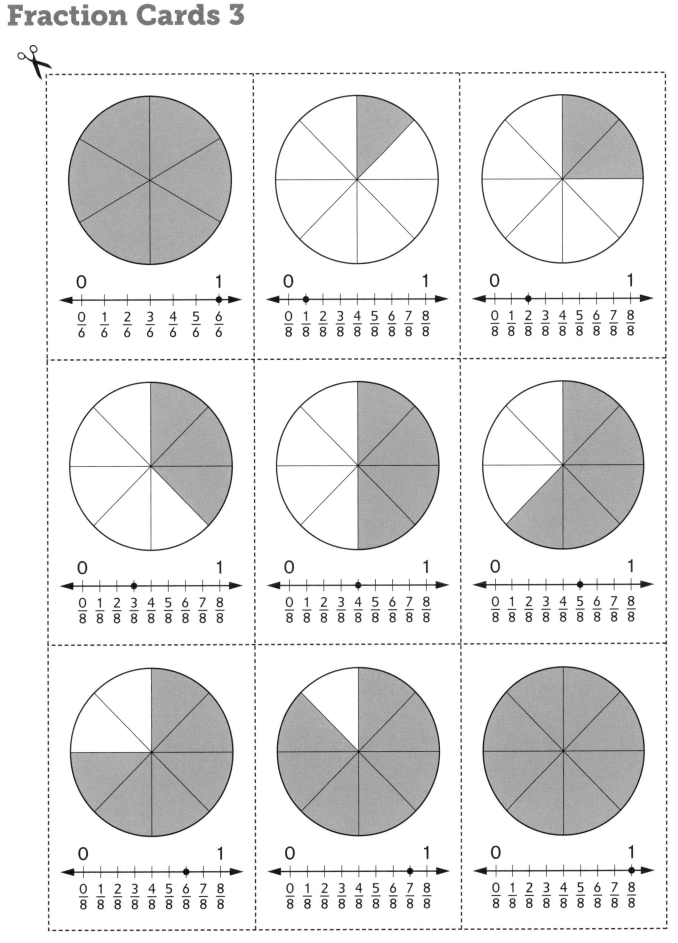

Fraction Cards 3 (continued)

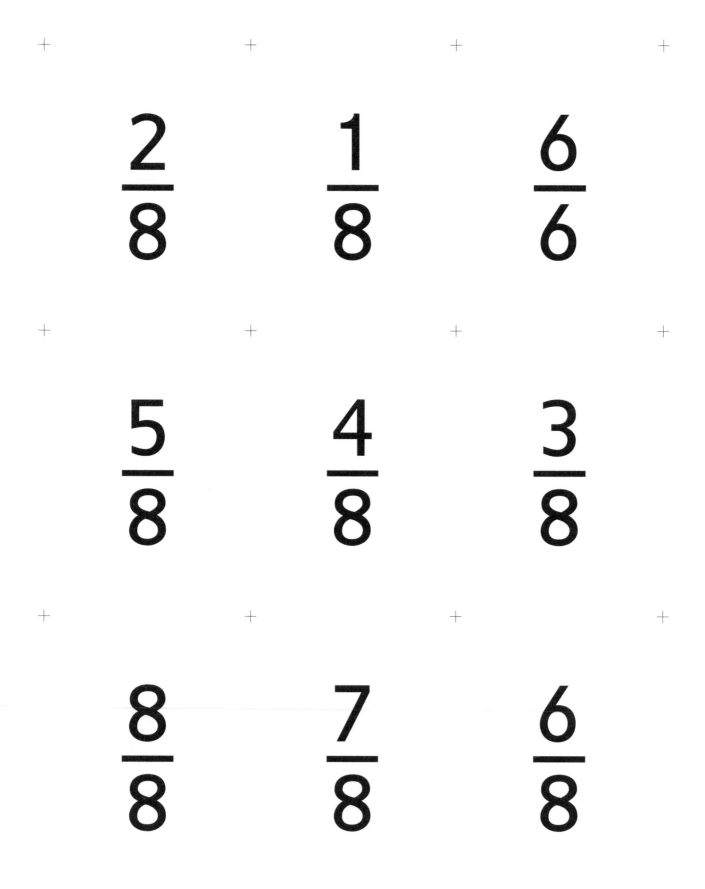

$$\frac{2}{8} \qquad \frac{1}{8} \qquad \frac{6}{6}$$

$$\frac{5}{8} \qquad \frac{4}{8} \qquad \frac{3}{8}$$

$$\frac{8}{8} \qquad \frac{7}{8} \qquad \frac{6}{8}$$

Back of Activity Sheet 18

Fraction Cards 4

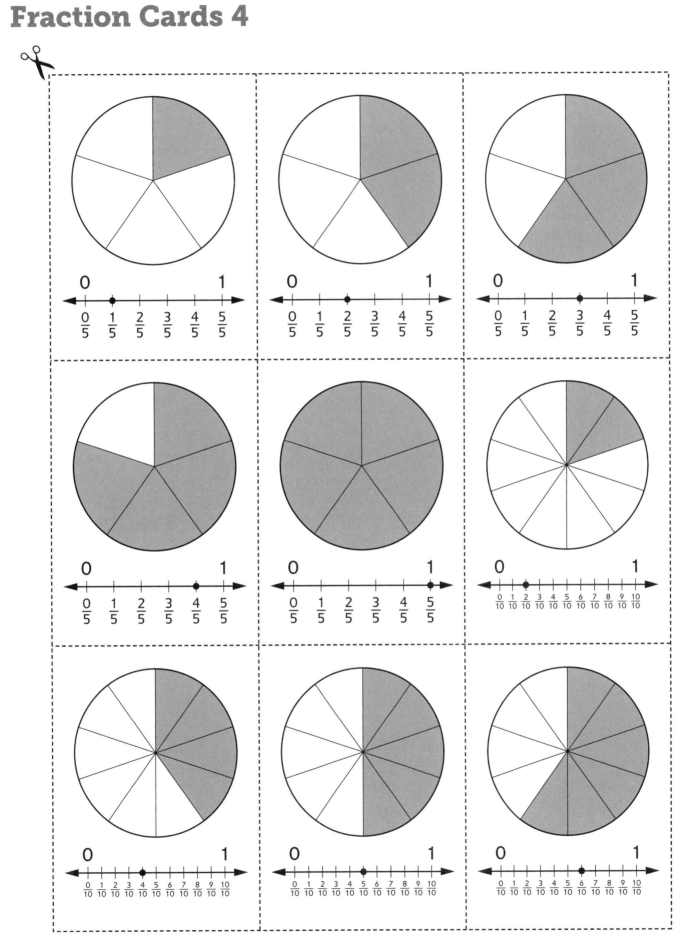

Fraction Cards 4 (continued)

$$\frac{3}{5} \qquad \frac{2}{5} \qquad \frac{1}{5}$$

$$\frac{2}{10} \qquad \frac{5}{5} \qquad \frac{4}{5}$$

$$\frac{6}{10} \qquad \frac{5}{10} \qquad \frac{4}{10}$$

Fraction Cards 5

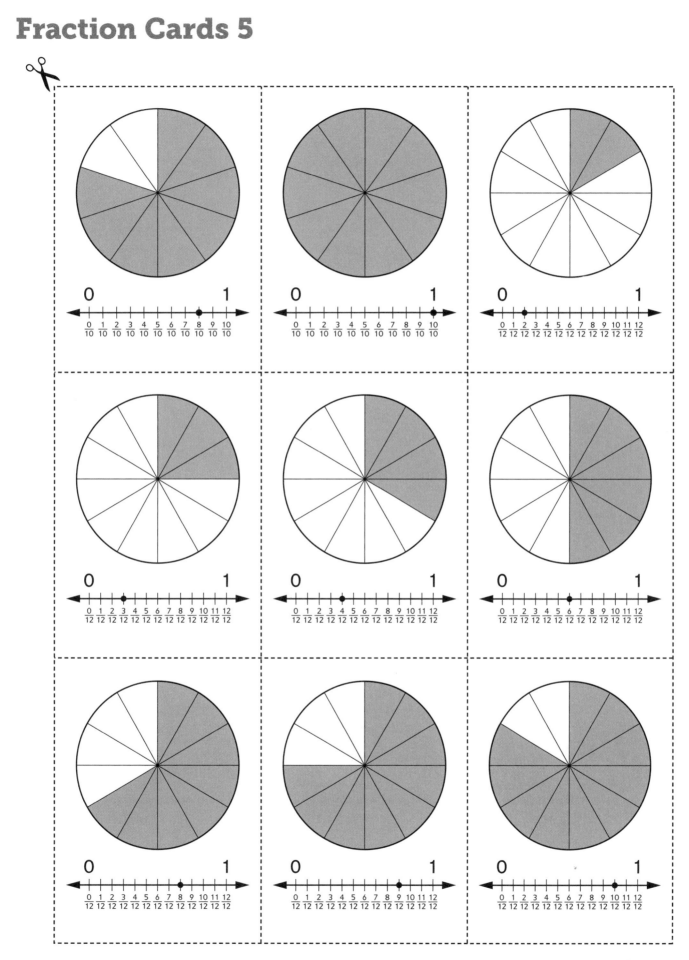

Fraction Cards 5 (continued)

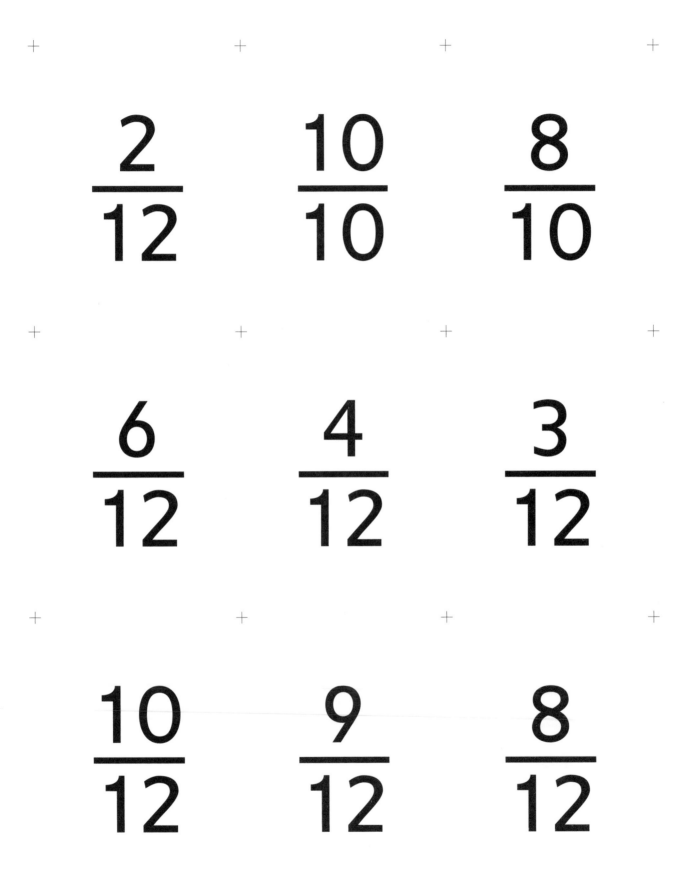

$$\frac{2}{12} \qquad \frac{10}{10} \qquad \frac{8}{10}$$

$$\frac{6}{12} \qquad \frac{4}{12} \qquad \frac{3}{12}$$

$$\frac{10}{12} \qquad \frac{9}{12} \qquad \frac{8}{12}$$

Back of Activity Sheet 20

Fraction Cards 6

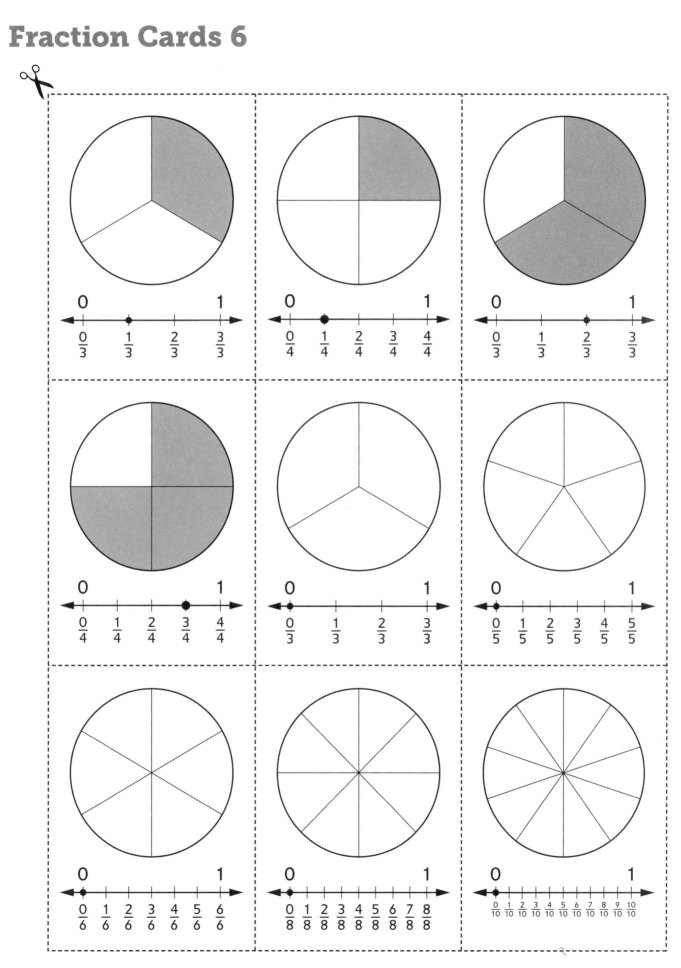

Fraction Cards 6 (continued)

$$\frac{2}{3}$$

$$\frac{1}{4}$$

$$\frac{1}{3}$$

$$\frac{0}{5}$$

$$\frac{0}{3}$$

$$\frac{3}{4}$$

$$\frac{0}{10}$$

$$\frac{0}{8}$$

$$\frac{0}{6}$$

Back of Activity Sheet 21

×, ÷ Fact Triangles: Remaining Facts

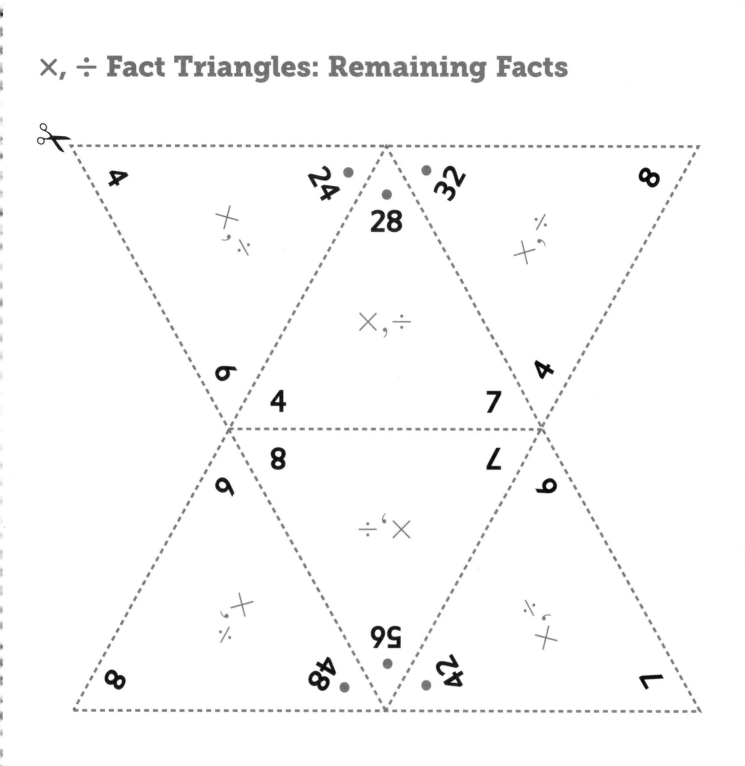